GC/WORKS/1
EDITION 3

The Government General Conditions of Contract for Building and Civil Engineering

VINCENT POWELL-SMITH
LLM, DLitt, FCIArb, MBAE

BSP PROFESSIONAL BOOKS
OXFORD LONDON EDINBURGH
BOSTON MELBOURNE

Copyright © Ingramlight Properties Ltd 1990
Note: GC/Works/1 — Edition 3
Standard Form of Contract is
Crown copyright

All rights reserved. No part of this
publication may be reproduced, stored
in a retrieval system, or transmitted,
in any form or by any means, electronic,
mechanical, photocopying, recording
or otherwise without the prior
permission of the copyright owner.

First published 1990

British Library
Cataloguing in Publication Data
Powell-Smith, Vincent
   GC/Works/1 – edition 3.
   1. Great Britain. Construction.
   Government contracts
   I. Title
   624

ISBN 0-632-02633-2

BSP Professional Books
A division of Blackwell Scientific
   Publications Ltd
Editorial Offices:
Osney Mead, Oxford OX2 0EL
   (Orders: Tel. 0865 240201)
25 John Street, London WC1N 2BL
23 Ainslie Place, Edinburgh EH3 6AJ
3 Cambridge Center, Suite 208,
   Cambridge, MA 02142, USA
54 University Street, Carlton,
   Victoria 3053, Australia

Set by DP Photosetting, Aylesbury, Bucks

Contents

Preface		vii
1	**Background to the form**	**1**
	Background and history	1
	Edition 3	3
	Main features	6
	Interpretation of GC/Works/1	8
	Contractual capacity of the Crown	9
	Form and content	11
2	**Formation of the contract**	**22**
	Introduction	22
	The tender and its acceptance	23
	Contract documentation	26
	The Limitation Act 1980	27
	Proposals for reform	30
3	**Contract documentation, information and staff**	**31**
	Introduction	31
	Clause 1 Definitions etc.	33
	Clause 2 Contract documents	38
	Clause 3 Bills of quantities	41
	Clause 4 Delegations and representatives	44
	Clause 5 Contractor's agent	46
	Clause 6 Contractor's employees	47
4	**General obligations**	**48**
	Introduction	48
	Clause 7 Conditions affecting works	57

Clause 8 Insurance	60
Clause 9 Setting out	62
Clause 10 Design	63
Clause 11 Statutory notices	65
Clause 12 Patents	66
Clause 13 Protection of works	67
Clause 14 Nuisance and pollution	68
Clause 15 Returns	69
Clause 16 Foundations	70
Clause 17 Covering work	71
Clause 18 Measurement	72
Clause 19 Loss or damage	73
Clause 20 Personal data	76
Clause 21 Defects	77
Clause 22 Government premises	79
Clause 23 Racial discrimination	80
Clause 24 Corruption	81
Clause 25 Records	82
5 Security	**83**
Introduction	83
Clause 26 Site admittance	84
Clause 27 Passes	85
Clause 28 Photographs	86
Clause 29 Secrecy	87
6 Materials and workmanship	**88**
Introduction	88
Clause 30 Vesting	90
Clause 31 Quality	94
Clause 32 Excavations	99
7 Commencement, programme, delays and completion	**101**
Introduction	101
Clause 33 Programme	103
Clause 34 Commencement and completion	105
Clause 35 Progress meetings	109
Clause 36 Extensions of time	111
Clause 37 Early possession	117
Clause 38 Acceleration	119
Clause 39 Certifying work	121

8	**Instructions and payment**	**122**
	Introduction	122
	Clause 40 PM's instructions	124
	Clause 41 Valuation of instructions — principles	135
	Clause 42 Valuation of variation instructions	136
	Clause 43 Valuation of other instructions	140
	Clause 44 Labour tax	144
	Clause 45 VAT	144
	Clause 46 Prolongation and disruption	145
	Clause 47 Finance charges	153
	Clause 48 Advances on account	156
	Clause 49 Final account	162
	Clause 50 Certifying payments	163
	Clause 51 Recovery of sums	164
	Clause 52 Cost savings	165
9	**Particular powers and remedies**	**166**
	Introduction	166
	Clause 53 Non-compliance with instructions	167
	Clause 54 Emergency work	168
	Clause 55 Liquidated damages	169
	Clause 56 Determination	172
	Clause 57 Consequences of determination for default	176
	Clause 58 Consequences of other determination	177
	Clause 59 Adjudication	179
	Clause 60 Arbitration	183
10	**Assignment, sub-letting, sub-contracting, suppliers and others**	**186**
	Introduction	186
	Clause 61 Assignment	187
	Clause 62 Sub-letting	188
	Clause 63 Nomination	191
	Clause 64 Provisional sums	195
	Clause 65 Other works	196
	Table of cases	**197**
	Table of statutes	**203**
	Index	**204**

Dedication

Semper in hoc libro vivat
carissima coniunx

Preface

The production of a new form of construction contract by the Property Services Agency of the Department of the Environment is an important event for all construction professionals. GC/Works/1 — Edition 3 was published in December 1989 and is the worthy successor of a long line of Government forms of building contract, the last of which, Edition 2 of GC/Works/1, will continue to be used alongside the new form for an interim period.

The sooner Edition 2 is formally withdrawn the better it will be for the industry because Edition 3 is vastly superior in every respect. It is a contract for the present decade and beyond and is certain to be widely used, not only by the PSA and other Government agencies (for whom it is intended) but also by those local authorities and private employers who favour the firm control which the conditions give. At all events, users and potential users need to become familiar with its provisions, many of which are innovatory, and to appreciate the way in which the new contract re-allocates risks.

My long involvement in the construction industry led to me having access to the various consultative drafts of what is now Edition 3 and I was also able to submit some comment on them. I have long been known to favour the Government forms of contract as being fair but firm and, under the PSA at least, fairly and competently administered. GC/Works/1 is, from a legal point of view, infinitely to be preferred to any of the 'negotiated' standard forms of contract which, inevitably, are compromise documents and display the inherent weaknesses and uncertainties resulting from 'committee drafting'. GC/Works/1 differs entirely in its approach to risk allocation from those negotiated forms and, in its third edition, is drafted in plain English.

Its clear draftsmanship and comparative lack of ambiguities have made my task in writing this book comparatively simple. It is not intended to be a legal textbook, but rather a guide for those in the construction industry who will be using the new form. For this reason, I have sometimes taken a clear view of some of the contract provisions where, in truth, there might be reasonable dispute amongst lawyers.

What I have written must be taken as the opinion of only one construction industry lawyer. At the end of the day, only Her Majesty's judges can say what meaning is to be attributed to the terms of any contract, but there have been mercifully few cases on earlier versions of what is now GC/Works/1, Edition 3. There is every reason to believe that this will continue to be the case. I do believe that GC/Works/1 is a good form of contract, and I hope that the industry will study it impartially and not brand it as 'unfair' or 'one-sided' — which it is not.

My original intention was to write a clause-by-clause analysis of the form with the contract text on one page and the commentary on the facing page. Unfortunately, Her Majesty's Stationery Office (in whom the copyright is vested) refused permission for the reproduction of the full text on the grounds that this would adversely affect sales of the contract form itself!

In the event, therefore, I have adopted the same approach as in some of my earlier commentaries on building contract forms and which, judging from sales at least, has been found acceptable to users.

Thus, after two brief introductory chapters, I opt for the clause-by-clause approach, but without reproducing the full text of the contract form itself. Readers should, therefore, obtain a copy of GC/Works/1, Edition 3, from Her Majesty's Stationery Office if they are to obtain the full benefit of the commentary. It is what the contract says that is decisive and not what people think it says.

Once again, I am grateful to my friend and fellow author David Chappell, MA, PhD, RIBA, who has been good enough to let me reproduce a set of his now well-known tabular summaries of certain important aspects of GC/Works/1 and which were produced for the purposes of a joint article of ours in the *Architect's Journal*. He has also been kind enough to produce the eight flow charts which enhance the text.

I am also very grateful to John Garnett, FRICS, MCIOB, Head of Claims at PSA, who has given me permission to reproduce some of his Stage Payment Charts by way of illustration with the *caveat* that, at the time of writing, the PSA has yet to decide between charts using weeks on a vertical axis and a more simplified version with percentage progress. I am equally grateful to Bruce Perry, FRICS of PSA Contracts Directorate for his help and advice. I should like to make it clear, however, that neither of these last two gentlemen nor the PSA is in any way responsible for any of the opinions expressed in this book. The views expressed are entirely my own, as are all errors of commission, interpretation or omission.

Finally, Julia Burden, Commissioning Editor at Blackwell Scientific Publications, has once again been more helpful than was called for. I am especially grateful to her for encouraging me during planning and production of this book as well as for seeing it through the press with commendable speed at a time when I was working in the Far East.

Funchais
Portugal
January 1990   VINCENT POWELL-SMITH

Chapter 1

Background to the form

Background and history

GC/Works/1, Edition 3, was published on 18 December 1989 and is formally titled *The General Conditions of Contract for Building and Civil Engineering*. It is an evolution of earlier Government forms of contract, the first of which, Form CCC/Works/1, was published in 1943 and ran into nine editions, the last being issued in 1959.

Form CCC/Works/1 was the work of a single draftsman whose brief was to draft a form which would enable vital national war-time projects such as the construction of fortifications and airfields to be completed without delay. At that time, the various ministries had their own separate conditions of contract, none of which were adequate to deal with the problems of the time, and the only other readily available standard form of contract — the so-called 'RIBA Standard Form of Building Contract' in its 1939 edition — provided contractors with too many opportunities to challenge the employer's actions during the currency of the project and thus delay or frustrate completion.

Accordingly, Form CCC/Works/1 gave strong powers to the Authority (the employer) and conferred extensive powers on the superintending officer, as he was then called, to issue directions and to settle disputed matters arising during the construction period on a 'final and conclusive' basis. A commentary on the 1970 reprint of CCC/Wks/1 is to be found in *Further Building and Engineering Standard Forms* by I.N.Duncan Wallace QC (1973, Sweet & Maxwell, London) and its draftsmanship evidently stood the test of time since the learned author described it as 'on the whole very good, combining both greater simplicity and precision' than other standard forms. He added that 'the doubtful policies and unnecessary complication which so often permit the advancement of unmerited claims by contractors under the professional forms are largely absent': *ibid.*, p. 3.

In 1959 *The General Conditions of Government Contract for Building and Civil Engineering Works* was published and given the reference number GC/Works/1. A second edition was published in September 1977 and has been revised from time to time, the last occasion being on 17 November 1987 when amongst other things, a new Condition 2A was introduced allowing the contract sum to be adjusted if, during the execution of the Works, the contractor encountered unforeseeable ground conditions certified as such by the superintending officer and which affected the content and carrying out of work under the contract. The author's *The General Conditions of Government Contract for Building and Civil Engineering Works* (1984, Business Press International, Sutton) may be referred to by those desiring a commentary on Edition 2.

The post-war revisions in fact modified what were sometimes thought of as the somewhat draconian contract terms necessitated by the country's war effort and restored in most respects what contractors regarded as their normal commercial rights.

Indeed, by the beginning of the last decade it was found that the financial completion of projects was taking longer and longer, largely as a result of contractors advancing numerous claims for prolongation and disruption, whether as a result of ordered variations or otherwise, and more and more disputes were arising. It is, however, a tribute to the original draftsmanship of the form (which largely remained unchanged) that few disputes reached either arbitration or the courts.

The Property Services Agency (PSA) of the Department of the Environment is the principal Government department contracting on Form GC/Works/1, and has been responsible for the drafting of Edition 3. In 1984 the PSA set up a working group to produce a new form of contract. Its membership was drawn from the Agency's Contracts, Quantity Surveying and Legal Directorate and it produced a radically different form of contract which was issued in August 1985 as a consultative draft of proposed Edition 3.

Amongst the most significant changes in that draft were proposals for dealing with prolongation and disruption claims and the valuation of variations by means of standard reimbursement provisions. The first of these was a fixed daily rate (liquidated prolongation costs) to reimburse the contractor for extra costs incurred by delays caused by variations and other specified reasons. The second was a fixed percentage addition to valuations of extra work (additional variation percentage) for the cost of

Background to the form

all disruption caused by variation instructions. Under the proposals, tenderers would have been required to specify the amounts of both fixed daily rates which they would require, and cost-based claims would have been allowable only for disruption or prolongation caused by other instructions.

As a result of objections to these and other radical proposals from the contractors' organisations — notably the Building Employers' Confederation (BEC) — the consultative draft was withdrawn and the Agency set up another working group which is responsible for the production of GC/Works/1, Edition 3 as published.

The second working group decided to produce a new contract, written in plain English, dealing with the problems which were known to arise under Edition 2 and which would meet seven objectives. These were to:

- Review the allocation of risk under the contract.
- Speed up claims settlement and the agreement of final accounts.
- Deal with the problems of finance charges consequent on case law development.
- Tackle the problems of contractor's liability for design.
- Provide incentives for contractors.
- Deal with the continuing problems caused by nominated sub-contractors.
- Incorporate the role of the Project Manager.

The working party set about its task with urgency and quickly produced an initial draft of the proposed Edition 3. This was circulated to interested bodies for comment. This document abandoned the earlier proposals for dealing with the problem of contractors' financial claims and was quite well received. It was, in effect, a totally new contract and not merely an amended version of Edition 2. A second draft, which took account of the comments received from the industry's representative bodies and other interested parties, was issued in 1989. There were not inconsiderable differences between the two consultative drafts. The final version was published by Her Majesty's Stationery Office on behalf of the Department of the Environment in December 1989 and again differs in points of detail from the second consultative draft.

Edition 3

Edition 3 of GC/Works/1 is intended for use with major works of new construction whether of building or civil engineering. At present, it is only available in a lump sum with quantities version, but the PSA intends to produce separate versions to give two types of lump sum contract (specification and drawings only, or supported by a firm bill of quantities) and two types of measure and value contract (bill of approximate quantities or schedule of rates). The Agency is also giving consideration to the supplementary conditions which will be needed to facilitate a full design and construct approach.

There is an interval between publication and the conditions coming into use. The exact phasing of transition from Edition 2 to Edition 3 will be a matter for the PSA businesses to decide but the earliest date for use is 1 April 1990. At the time of writing, Edition 2 has not been withdrawn formally

Edition 3 is described as an evolution of Edition 2, but this is true only in part (see Table 1.1). Some of the former provisions are unchanged in principle though they have been redrafted. However, Edition 3 is

Table 1.1: Edition 2 and Edition 3 clause comparison

Edition 3 clause number	Heading in Edition 3	Edition 2 clause number
	Contract Documentation, Information and Staff	
1	Definitions etc.	1
2	Contract documents	4
3	Bills of quantities	5
4	Delegations and representatives	16
5	Contractor's agent	33
6	Contractor's employees	36
	General Obligations	
7	Conditions affecting works	2
8	Insurance	—
9	Setting out	12
10	Design	—
11	Statutory notices	14
12	Patents	15
13	Protection of works	17
14	Nuisance and pollution	18
15	Returns	34
16	Foundations	21
17	Covering work	22
18	Measurement	37
19	Loss or damage	26
20	Personal data	—
21	Defects	32
22	Government premises	35
23	Racial discrimination	52
24	Corruption	55
25	Records	—

Background to the form

Edition 3 clause number	Heading in Edition 3	Edition 2 clause number
	Security	
26	Site admittance	56
27	Passes	57
28	Photographs	58
29	Secrecy	59
	Materials and Workmanship	
30	Vesting	3
31	Quality	13
32	Excavations	20
	Commencement, Programme, Delays and Completion	
33	Programme	—
34	Commencement and completion	6
35	Progress meetings	—
36	Extensions of time	28
37	Early possession	28A
38	Acceleration	—
39	Certifying work	42
	Instructions and Payment	
40	PM's instructions	7
41	Valuation of instructions — Principles	—
42	Valuation of variation instructions	9
43	Valuation of other instructions	—
44	Labour tax	119
45	VAT	—
46	Prolongation and disruption	53
47	Finance charges	—
48	Advances on account	40
49	Final account	41
50	Certifying payments	42
51	Recovery of sums	43
52	Cost savings	—
	Particular Powers and Remedies	
53	Non-compliance with instructions	8
54	Emergency work	49
55	Liquidated damages	29
56	Determination	44
57	Consequences of determination for default	45
58	Consequences of other determination	46
59	Adjudication	—
60	Arbitration	61
	Assignment, Sub-letting, Sub-contracting, Suppliers and Others	
61	Assignment	27
62	Sub-letting	30
63	Nomination	31
64	Provisional sums	38
65	Other works	50

thoroughly innovative and contains many distinguishing features and is effectively a completely fresh contract.

Written in plain English, it is undoubtedly the best standard form of building contract available and shows all the advantages of unilateral provenance. Since it is drafted on behalf of the Department of the Environment it is more clear cut than the negotiated forms such as the JCT Standard Form JCT 80 and the Intermediate Form IFC 84 which are often an unhappy compromise. Edition 3 also approaches the allocation of risk in a sensible and businesslike way.

Edition 3 of GC/Works/1 is 'project orientated' The guiding principle has been to frame terms which are efficacious and foster quick decisions. The philosophy of the British Property Federation (BPF) system of building procurement has clearly influenced the draftsmen since they have adopted many of the features of that excellent system which are also found in the Association of Consultant Architects' form (ACA 2), the second edition of which was published in 1984. The BPF system for building design and construction is fully described in *The Manual of the BPF System* (1983, British Property Federation, London).

Some of the special features of GC/Works/1 — such as the unilateral power of determination conferred on the the Authority by clause 56 — result from the special circumstances of many Government projects.

Main features

The distinguishing features and more important or innovatory conditions of Edition 3 are:

- *A Programme.* This is a new concept, defined in general terms in clause 1 and with the detailed provisions set out in clause 33. These require the programme to be a well-considered document of considerable depth and sophistication.
- *A Project Manager* (PM) who is charged with the employer's day-to-day role and superintendance on his behalf. He could well be an architect or an engineer. The PM can act through duly appointed representatives, and is given extensive powers of delegation.
- *A measure of design liability is imposed on the contractor.* Clause 10 recognises that in most projects it is impossible for the design to have been fully specified at tender stage down to the last nut and bolt and that any competent contractor will at least carry out some on-site secondary design. Many decisions taken on site by building craftsmen are in fact 'design decisions'. Design or other drawings

must be approved by the PM. His approval does not lessen the contractor's liability.
- *Effective power to deal with defects during the maintenance period.* Clause 21 provides that defects notified to the contractor which the employer *believes* to be due to the contractor's default must be corrected by the contractor. If the contractor alleges that the employer's judgment is incorrect, he must pursue his argument after completion of the remedial works. If his argument is accepted then he will be paid for the remedial works.
- *Monthly progress meetings.* Before each meeting the contractor must submit comprehensive reports to the PM who must in turn issue a statement immediately after each meeting setting out key decisions and status reports.
- *Clear provisions for interim and final extensions of time.* The contractor has no right to any extension of time for bad weather. The contractor must accept the risk that the actual weather will be better or worse than that he has allowed for when tendering.
- *Acceleration can be requested by the employer or proposed by the contractor.* This depends on agreement and the cost implications are dealt with as part of the acceleration agreement between the parties.
- *Clear-cut principles for valuing instructions.* The valuation includes prolongation or disruption of both varied and unvaried work. There is a basic division of valuation procedures between variation instructions (VIs) and other instructions and an optional lump sum quotation system for variations. This is based on the ACA/BPF model.
- *Limited provisions for prolongation and disruption costs.* Reimbursement is based on 'expense' and limited finance charges are payable automatically on certain payments due to the contractor but are not otherwise recoverable.
- *Entirely new payment provisions.* Payments are by reference to stage payment charts given to the contractor with the tender documents.
- *Tight final account procedures* with strict time limits on both sides.
- *Cost saving proposals can be made by the contractor.* If accepted, there is equal sharing of cost savings made.
- *Adjudication on disputes during contract progress.* Only the contractor can request it and there is a preliminary three-month cooling-off period before the procedure can be invoked.
- The treatment of *nominated sub-contractors* differs from JCT 80 and most other standard form contracts. The contractor is in principle

fully reponsible for them. If a nominated sub-contractor fails, the employer pays only the sum that would have been paid had there been no failure. The important new exception to this rule is that the employer reimburses the contractor for any costs incurred where failure of the original nominee is due to his insolvency and where, despite his best endeavours, the contractor has failed to recover from the orginal sub-contractor in receivership or liquidation.
- *No contractual provision entitling the contractor to terminate the contract.* He must rely on his common law rights. In contrast, the employer is given express power to determine not only for the contractor's default, but also without reason.
- *Strict time limits* are specified though, except in relation to the grant of extensions of time, they can be extended by agreement (clause 1(4)).

Interpretation of GC/Works/1

Since GC/Works/1 is a form of unilateral provenance and not a document which is prepared jointly by bodies including the representatives of employers and contractors, as are the Joint Contracts Tribunal's standard forms, any ambiguities in it will be construed *contra proferentem* the employer, that is against the party putting the form forward.

The *contra proferentem* rule is an interpretative rule used by the courts when interpreting a contract and has been expressed in this way:

'If there is an ambiguity in a document which all the other methods of construction have failed to resolve so that there are two alternative meanings to certain words the court may construe the words against the party seeking to rely on them and give effect the the meaning more favourable to the other party': D. Keating, *Building Contracts*, 4th edition, p.34.

In appropriate circumstances GC/Works/1 could attract the application of the Unfair Contract Terms Act 1977 since it could be regarded as being the employer's 'written standard terms of business' for the purposes of section 3 of the 1977 Act. Section 3 provides, so far as material:

'(1) This section applies as between contracting parties where one of them deals . . . . on the other's written standard terms of business.
(2) As against that party, the other cannot by reference to any contract term —

(a) when himself in breach of contract, exclude or restrict any liability of his in respect of the breach; or
(b) claim to be entitled —
   (i) to render a contractual performance substantially different from that which was reasonably expected of him, or
   (ii) in respect of the whole or any part of his contractual obligation to render no performance at all, except in so far as .... the contract term satisfies the requirement of reasonableness.'

The test of 'reasonableness' is contained partly in section 11 and partly in Schedule 2 which sets out some guidelines. Sections 2 to 7 apply only to business liability. There is no definition of 'business' in the Act, but section 14 provides that the term includes 'a profession and the activities of any Government department or a local or public authority'.

It is the application of the *contra proferentem* rule that is the most important legal consequence of the fact that GC/Works/1 is drafted principally with the Government's interests in mind. However, this may be of more academic than practical interest since, like its predecessors, Edition 3 is relatively free from the the ambiguities and obscurities which characterise the negotiated forms of contract.

Contractual capacity of the Crown

GC/Works/1 is intended primarily for use by Government departments which, technically, are emanations of the Crown. In general, the Crown has the same power to make contracts as local authorities, companies in the private sector or individuals, although contracts made with Government departments are affected by certain rules which arise from the fact that such contracts are subject to political considerations. This is a complex subject and outside the scope of this book, but the following points must be made:

- ***There are limits on the contractual capacity of the Crown***

The Crown enjoys a capacity to contract at common law. It is also given power to make contracts for particular purposes by statute. Before the Crown Proceedings Act 1947 the procedure for suing the Crown was extremely cumbersome and, despite the changes made by the Act, certain limitations on the power of the Crown to bind itself by contract have survived, although their extent is not entirely clear.

It is, for example, the law, that:

'It is not competent for the Government to fetter its future executive action, which must necessarily be determined by the needs of the community when the question arises. It cannot by contract hamper its freedom of action in matters which concern the welfare of the State.'

This wide rule was laid down by Mr Justice Rowlatt in *Rederiaktiebolaget Amphitrite* v. *The King* (1921) and its scope is uncertain. In *Robertson* v. *Minister of Pensions* (1949) Lord Justice Denning (as he then was) suggested that the rule does not apply to ordinary commercial contracts such as building contracts. Although there are judicial suggestions that it does, this view probably represents the correct position.

- *There is a distinction between the Crown and public corporations*

As a general principle, it may be said that Government departments, such as the Department of the Environment of which the PSA forms part, are regarded as an emanation of the Crown. As such, they are subject to any special rules about the Crown's contractual capacity. Statutory public corporations are not emanations of the Crown and in practice they are given the widest possible contractual and other powers by the statute which created them. A public corporation may, in fact, be acting on behalf of the Crown. In principle it would then be the Crown which was bound.

- *Crown contracts must necessarily be made through the agency of human beings*

The basic rule of agency is that an agent has authority to bind his principal either when he has *actual authority* to do so or when he has *ostensible authority*, i.e., when he appears to the outside world to possess authority to enter into contracts of the type in question. However, if a servant or agent of the Crown assumes an authority which he does not in fact possess and enters into an unauthorized contract, it was held in *Attorney-General for Ceylon* v. *Silva* (1953) that the Crown will not be bound by his act unless it has held him out to have that authority.

So far as actions during the currency of a contract on GC/Works/1, Edition 3 terms are concerned, the matter is dealt with by clause 4(1):

'Any decision to be made by the Authority under the Contract may be made by any person or persons authorised to act for him for that purpose. The PM shall be deemed authorised to act generally for the Authority *subject to any exclusions set out in the Abstract of Particulars*': [Emphasis supplied].

Background to the form

In practice, although the Project Manager is the Authority's general representative, he will not exercise the totality of the Authority's powers. It is to be expected that the abstract of particulars may well reserve to the Authority such matters as determination of the contract, payment and retention and questions arising therefrom and the exercise of the Authority's right of set-off.

- *Crown contracting procedures*

The contracting procedures of Government departments have evolved over the years as a result of the general oversight of the Treasury and following reviews of particular contracts by the Public Accounts Committee. As a result, a number of general principles for Government contracting have been adopted. The main principles are:

(i) Whenever reasonably practicable, Government contracts should be let by competitive tender.
(ii) The financial and commercial standing of contractors should be verified before contracts are awarded to them, i.e., there are lists of approved contractors.
(iii) In general, contract orders should not be placed without agreement as to price.

These principles are applied by the PSA when contracts are let under GC/Works/1.

Form and content

The printed conditions of contract are issued in the form of a 38 page booklet. There are no annexed articles of agreement. Articles of agreement incorporating the contract conditions must be entered into separately.

The printed document consists of 65 conditions (see Table 1.2) which are divided into eight un-numbered headings arranged according to subject matter. The sections are as follows:

- Contract documentation, information and staff (*conditions 1 to 6*)
- General obligations (*conditions 7 to 25*)
- Security (*conditions 26 to 29*)
- Materials and workmanship (*conditions 30 to 32*)
- Commencement, programme, delays and completion (*conditions 33 to 39*)

Table 1.2: GC/Works/1 — Edition 3: Conditions

Clause	Clause subject
	Contract Documentation, Information and Staff
1	Definitions etc
2	Contract documents
3	Bills of quantities
4	Delegations and representatives
5	Contractor's agent
6	Contractor's employees
	General Obligations
7	Conditions affecting works
8	Insurance
9	Setting out
10	Design
11	Statutory notices
12	Patents
13	Protection of works
14	Nuisance and pollution
15	Returns
16	Foundations
17	Covering work
18	Measurement
19	Loss or damage
20	Personal data
21	Defects
22	Government premises
23	Racial discrimination
24	Corruption
25	Records
	Security
26	Site admittance
27	Passes
28	Photographs
29	Secrecy
	Materials and Workmanship
30	Vesting
31	Quality
32	Excavations
	Commencement, Programme, Delays and Completion
33	Programme
34	Commencement and completion
35	Progress meetings
36	Extensions of time
37	Early possession
38	Acceleration
39	Certifying work

Clause	Clause subject
	Instructions and Payment
40	PM's instructions
41	Valuation of instructions — Principles
42	Valuation of variation instructions
43	Valuation of other instructions
44	Labour tax
45	VAT
46	Prolongation and disruption
47	Finance charges
48	Advances on account
49	Final account
50	Certifying payments
51	Recovery of sums
52	Cost savings
	Particular Powers and Remedies
53	Non-compliance with instructions
54	Emergency work
55	Liquidated damages
56	Determination
57	Consequences of determination for default
58	Consequences of other determination
59	Adjudication
60	Arbitration
	Assignment, Sub-letting, Sub-contracting, Suppliers and Others
61	Assignment
62	Sub-letting
63	Nomination
64	Provisional sums
65	Other works

- Instructions and payment (*conditions 40 to 52*)
- Particular powers and remedies (*conditions 53 to 60*)
- Assignment, sub-letting, sub-contractors, suppliers and others (*conditions 61 to 65*)

There is also a general note on the use of supplementary provisions. This states that special conditions may be incorporated by listing in the abstract of particulars if they are required. Since GC/Works/1 contains only a fluctuations provision of the most limited type (clause 44 — labour tax) conditions for price variation will be set out in the abstract of particulars as occasion requires.

The PSA has a number of standard model Supplementary Conditions. Those currently available are listed in Table 1.3.

Table 1.3: PSA Supplementary Conditions

- Articles and other products associated with a works contract covered by the contract relating to conditions dealing with the drawing rights, drawings, specifications, manufacturing data, use of the documents, information etc.
- Control of noise: Incorporation of BS 5228
- Crown rights in inventions and designs
- Design develop and construct contracts — single stage
- Design develop and construct contracts — two stage
- Dredging works
- Galleries and museums — special provisions about access, working methods etc.
- Geotechnical investigations
- Indemnifications against loss or damage caused by ammunition
- Mastic asphalt and built-up bitumen felt work
- Open-book accounting for certain costs
- Ownership of patents and registered designs
- Payment for goods and materials held off site
- Racial discrimination
- Security provisions for USAF bases
- Vesting of goods and materials not on site
- Variation of price. A range of supplementary conditions for general and specialist work on the basis of both formula and traditional methods of calculation
- Working in naval establishments

Table 1.4: Authority's powers under GC/Works/1 — Edition 3

Clause	Power	Comments
8(3)	Request copy of insurance effected or held.	Contractor must provide certificate from his insurer or broker within 21 days from acceptance of tender and within 21 days of later renewal.
8(4)	Effect appropriate insurance cover; deduct cost from advance due to contractor	If contractor fails to maintain.
19(2)	Agree to compensation for loss or damage suffered by the authority.	Instead of making good.
21(3)	Do anything necessary to make good defects notified to the contractor. Recover costs and expenses from the contractor.	If contractor fails to comply with clause 21.
24(2)	By notice determine the contract.	If the contractor: • is in breach of this clause 24; *or* • is convicted of offence under the Prevention of

Background to the form

Clause	Power	Comments
	Recover the value of such gift or consideration.	Corruption Acts 1889 to 1916. Without prejudice to powers conferred by clause 51. Decision is final and conclusive.
37(1)	Take possession of any part of the works.	Before completion if PM has certified it complete to his satisfaction and it is either: • a section; *or* • other part of the works agreed or subject of PM's instruction that possession is to be given before completion of the works. After the date of the certificate, the completed part is no longer part of the works for clauses 8 19 and 30.
42(3)	Agree the aggregate amount to which the contractor is entitled after acceptance of a quotation other than specified in the quotation.	For complying with the VI.
48(6)	Withhold payment to contractor of amount until PM is satisfied.	If PM not satisfied that amount due to nominated sub-contractor or supplier for 'things for incorporation', covered by any previous advance, has been paid.
49(1)	Estimate the final sum.	
50(3)	Decide a dispute about the contractor's right to a certificate or sums to be certified.	If so referred at the request of the contractor. The authority's decision is final and conclusive. Clause 50(3) does not apply to a dispute about the balance of the final sum due.
52(2)	Require further information relating to the proposal.	If the contractor submits a proposal to reduce the cost of the works.
53	Provide labour and/or 'things'; *or* enter into a contract for execution of work to give effect to an instruction.	If the contractor fails to comply with notice from PM requiring compliance with the instruction without prejudice to power to determine the contract.

Table 1.4: Authority's powers under GC/Works/1 — Edition 3 (cont.)

Clause	Power	Comments
54(2)	Make arrangements for emergency work to be carried out.	If the contractor fails to carry out the work promptly. The contractor must reimburse any extra costs.
55(3)	Deduct liquidated damages from advances to which the contractor may be entitled under clause 48.	
55(5)	Waive right to deduct or recover liquidated damages.	
56(1)	Determine the contract by notice to the contractor.	Without prejudice to any other power of determination.
56(3)	Give directions in relation to the performance of any work etc.	The decision is final and conclusive.
56(6)(e)	Determine whether failure by the contractor to comply with clause 25 is prejudicial to the interests of the Crown.	The decision is final and conclusive.
57(1)(b)	Hire any person. Employ other contractors. Use any 'things' on site. Purchase or do anything necessary for completion of the works. Employ other persons to complete the works.	If the authority determines the contract under clause 56.
57(1)(d)	Pay to any sub-contractor or supplier any amount due.	If the authority determines the contract under clause 56. If the PM certifies the amount as included as included in any previous advance to the contractor.
57(2)(c)	Elect to keep 'things for incorporation' brought on to site or in course of preparation off site.	
58(2)	Elect to keep 'things for incorporation' brought on to site or in course of preparation off site.	
61	Consent to the assignment of the contract.	In writing.
62(3)	Request the contractor to take steps to ensure that a sub-contractor performs all his obligations.	At any time.
63(2)	Direct the way in which a person may be nominated or appointed.	
63(2)	Order and pay for any part of prime cost items direct. Deduct such payments from the contract sum.	Less an amount in respect of contractor's profit at rate included in bills of quantities or schedule of rates adjusted pro-rata on the amount paid by the authority.

Background to the form

Clause	Power	Comments
63(5)	Require the contractor to supply detailed scheduled of rates properly used for calculating the contract sum or sub-contract sum.	If bills of quantities not provided in respect of any work or 'things' to which this clause applies.
63(7)	Agree to the contractor determining or novating a nominated sub-contract.	
65(1)	Execute other works on the site at the same time as the works are being executed.	At any time.

Table 1.5: Authority's duties under GC/Works/1 — Edition 3

Clause	Duties	Comments
3(3)	Rectify any error or omission in the bills of quantities in accordance with clause 42(4).	Contract sum shall be increased or decreased accordingly.
11(3)	Reimburse the amount of any fee or charge paid by the contractor.	If properly incurred in accord with this clause.
19(4)	Notify contractor of any claim or proceedings in respect of loss or damage.	As soon as possible.
19(5)	Reimburse the contractor for loss in taking clause 19(2) and 19(3) actions.	To the extent the loss was caused by • Authority's neglect or default *or* • Accepted risk or unforeseeable ground conditions *or* • Other circumstances outside contractor's or sub-contractor's control.
21(2)	Reimburse the contractor for any cost incurred to the extent the authority is satisfied defects were not caused by: • Contractor's default; *or* • Circumstances within contractor's control.	After completion of remedial works.
34(1)	Notify contractor that he may take possession of the site to commence the works.	Within the period in abstract of particulars.
37(6)	Pay to the contractor: • Half the share apportioned in accordance with clause 37(5). • Remaining half.	In respect of the completed part. When PM has certified after maintenance period that the completed part is in a satisfactory state.
38(1)	Direct the contractor to submit: • Priced proposals to achieve early completion; *or* • Why early completion cannot be achieved.	If the authority wishes to achieve early completion.
38(2)	Specify in notice to contractor: • The accelerated date.	If authority accepts proposals.

Table 1.5: Authority's duties under GC/Works/1 — Edition 3 (cont.)

Clause	Duty	Comments
	• Programme amendments. • Revised contract sum. • Revised stage payment chart. • Any other agreed relevant contract amendment.	
38(3)	Consider contractor's proposals for accelerated completion. Take clause 38(2) action.	If he accepts them.
45(2)	Add an appropriate sum in respect of tax.	When making payment due to the contractor.
45(4)	Recover from the contractor tax additional to what would have been paid if the contractor had carried out his duties under the contract.	If contractor fails to carry out duties under the contract and another contractor is engaged to do the duties and is paid to him.
47(1)	Pay contractor interest charges incurred because: • Authority, PM or QS has failed to comply with any time limit; or • QS varies a decision notified to the contractor.	
48(7)	Accumulate the balance of any sum which is less than 100% as a reserve.	As mentioned in clause 48(2).
49(1)	Pay the contractor the Authority's estimate of the final sum less half the reserve.	As soon as reasonably possible.
49(4)(a)	Pay excess to the contractor.	If final sum has been calculated before end of maintenance period and balance exceeds reserve which the Authority is entitled for time being to retain.
49(5)(a)	Pay excess to the contractor.	If final sum exceeds amount paid to contractor on an estimated basis.
56(2)	Specify which clause 61(6) grounds, if any, apply.	In a notice of determination.
57(2)	Hold the amount of the excess.	If the 'first amount' exceeds the 'second amount'.
58(2)	Pay the contractor the amount of the excess.	If the 'third amount' exceeds the 'second amount'.
58(4)	Make such allowance as in his opinion is reasonable.	If the contractor, of the opinion that determination losses have not been fully reimbursed, refers circumstances to the Authority.

Background to the form

Clause	Duty	Comments
63(7)	Nominate a replacement sub-contractor or supplier or direct the contractor to complete the work.	If the Authority has agreed to determination of sub-contract.
63(9)	Reimburse between extra cost of completing sub-contract work and cost contractor should have recovered from original sub-contractor.	If the nominated sub-contract is determined or assigned because sub-contractor is insolvent.

Table 1.6: Project Manager's duties under GC/Works/1 — Edition 3

Clause	Duty	Comments
2(5)	Provide to the contractor one copy of the contract drawings, specification and blank bills of quantities and drawings issued during the progress of the works.	Free of charge.
7(4)	Certify conditions as 'unforeseeable ground conditions'. Notify the contractor of his decision.	If the PM agrees that the ground conditions specified in clause 7(3) notice could not reasonably have been foreseen having regard to information the contractor should have had in accordance with clauses 7(1) and 7(3).
9(1)	Provide such dimensional drawings, levels etc. as he considers reasonably necessary to enable the contractor to set out.	The contractor must set out, provide the equipment and be solely responsible for setting out.
27	Issue passes to the contractor.	Where workpeople are required to have a pass before admission to site. They must be returned on demand.
35(2)	Specify time and place of progress meetings.	At intervals of one month subject to contrary instruction.
35(4)	Give the contractor a statement specifying: • Extent to which project is one time, delayed or early; *and* • Matters delaying or likely to delay completion; *and* • Steps agreed or appropriate to reduce or eliminate effects of delay; *and* • Applications for and extensions of time under clause 36.	Within seven days of each progress meeting.

Table 1.6: Project Manager's duties under GC/Works/1 — Edition 3 (cont.)

Clause	Duty	Comments
36(1)	Notify contractor regarding extension of time for completion of the works or section.	If he has received a written request from the contractor or he thinks there is or is likely to be delay preventing completion by the relevant date. Within six weeks of notice.
36(2)	Award an extension only if satisfied delay is due to: • Execution of varied work; *or* • Act, neglect or default of the authority or PM; *or* • Strike or lock-out; *or* • Accepted risk or unforeseeable ground conditions; *or* • Other circumstances wholly beyond the contractor's control (not weather conditions).	
36(3)	Indicate whether his decision is interim or final. Keep interim decisions under review until he can give final decision.	
36(5)	Come to a final decision on all interim extensions within 21 days of completion of the works.	In so doing he is not entitled to reduce interim extensions already awarded.
36(5)	Notify the contractor whether in the light of his claim the decision has been amended.	Within 28 days of the receipt of a claim from the contractor disputing a decision and specifying grounds entitling him to more extension.
37(3)	Certify the value of the completed part of the works in order to calculate the reduction of liquidated damages and to adjust the amount of reserve.	If the PM certifies that part of the works has been completed to his satisfaction. A certificate under this clause must be issued as soon as possible thereafter.
37(4)	Determine the value of the completed part.	
37(5)	Apportion the reserve accumulated in accordance with clause 48 so that the ratio between reserve to completed part and reserve of the whole is the same as the value of the completed part to the value of the whole.	With effect from the date of the clause 37(1) certificate.
39(1)	Certify the date on which the works or section are completed to his satisfaction. Issue a certificate when the works are in a satisfactory state.	After the end of the longest relevant maintenance period.

Clause	Duty	Comments
42(3)	Notify the contractor whether or not a lump sum quotation has been accepted.	Not later than 21 days from receipt.
48(3)	Amend the stage payment chart.	If he has recorded in a statement after a progress meeting that the works are in delay or ahead of programme.
48(4)	Make appropriate amendments to the stage payment chart.	Immediately the authority has accepted a proposal under clause 52.
50(1)	Certify sums to which the contractor is entitled under clauses 48 and 49.	
52(4)(b)	Issue any necessary agreed extension of time.	If the authority accepts a written proposal from the contractor under clause 52(1).
57(1)(e)	Certify the cost to the authority of completion of the works.	If the authority determines the contract and after the QS has ascertained the cost.

Tables 1.4 and 1.5 set out the Authority's powers and duties under Edition 3, while Table 1.6 lists the Project Manager's duties.

Chapter 2
Formation of the contract

Introduction

Users inevitably refer to the printed conditions themselves as 'the contract', but the conditions alone do not constitute the contract since they say nothing at all about the work to be done and are silent on many other important matters. This is in fact emphasised by clause 1(1) which defines 'the Contract' as meaning the conditions of contract, abstract of particulars, specification, drawings, bills of quantities, the tender and the Authority's written acceptance thereof. It is always necessary to consider the specific agreement the parties have made and to interpret this in light of the general law of contract.

GC/Works/1 must therefore be read and interpreted against the background of the common law. In particular, implied terms may be read into it in order to make the contract commercially effective: *The Moorcock* (1889). It may well be, as Lord Esher MR remarked in *Lynch* v. *Thorne* (1956), that:

> 'Where there is a written contract expressly setting forth the bargain between the parties it is, as a general rule, also well established that you only imply terms under the necessity of some compulsion.'

Nonetheless, even where there is a comprehensive written contract and an apparently complete bargain, the courts have been willing to add a term on the grounds that without it the contract will not work. Terms of this sort are written in by the courts on the basis of the presumed intention of the parties, and in ordinary building contracts terms of co-operation on the part of the employer, and against hindrance or prevention by him may be implied on this basis: *London Borough of Merton* v. *Stanley Hugh Leach Ltd* (1985).

That the doctrine of the implied term in building contracts is alive and kicking is shown by the recent case of *Bruno Zornow (Builders) Ltd* v. *Beechcroft Developments Ltd* (1989) where a dispute about liquidated damages arose out of a substantially amended JCT contract, 1963 edition, on extremely complex facts. The contract was varied by agreement between the parties so as to substantially increase its scope by the inclusion of additional work. Judge John Davies QC, Official Referee, held that, in order to give business efficacy to the contract, it was necessary to imply into it a term giving dates for completion tied to a provision for liquidated damages.

The question of implied terms is fully dealt with in all the standard texts on general contract law such as Cheshire, Fifoot and Furmston's *Law of Contract,* 11th edition (1986, Butterworths, London) pp.126–140.

The important point is that the law of building contracts is fixed firmly in the mainstream of the general law and neither GC/Works/1 nor any other standard form of contract is a self-sufficient code containing all the contractual and other obligations of the parties. The relationships and rights between the parties are also affected by various statutes, such as the Limitation Act 1980, as amended, and the Misrepresentation Act 1967, for example.

The making of the contract between the Authority and the contractor is governed by the ordinary rules of contract law. English law has no special rules about the way in which building and civil engineering contracts must be made. In practice, the parties normally adopt a formal approach, whether the project is large or small, and this is the sensible course to follow.

The tender and its acceptance

Simple contracts, i.e., those which are not made under seal as deeds, are created by a process of offer and acceptance. All that is required is that there should be an unequivocal offer made by one party to the other and an unconditional acceptance of the offer by the other party. The agreement thus arrived at must be supported by 'consideration', e.g., the price. If these elements are present, no further formalities are required, although this analysis is a simplistic one and reference should be made to the standard texts on contracts such as John Parris, *The Making of Commercial Agreements* (BSP Professional Books, London, 1988).

The PSA's usual tendering procedure is to invite tenders from

**FORM of TENDER FOR USE WITH GC/WORKS/1 CONDITIONS
(FIRM BQ's)**

(NOTE: This form must be used for submitting a tender. Photocopies will not be accepted.)

To: the Secretary of State for the Environment

....

....

....

Sir

1. This Tender is returned in response to your invitation dated ....

   ref .... to execute works at ....

   comprising ....

   .... and fully described or shown in the specification, drawings and Bills of Quantities.

2. The following documents shall be part of the Contract if this Tender is accepted:-

   (1) General Conditions of Government Contracts for Building and Civil Engineering works, Form GC/Works/1 (Edition 3), plus amendment No ....

   (2) * Abstract of particulars (form C1009 Abs).

   (3) Addendum to Abstract of Particulars (form C1009 Abs-Add).

   (4) * Supplementary Condition(s) No ....

   ....

   (5) * Composite Specification(s), Particular Specification.

   (6) Standard Fire Precautions Booklet P5.

   (7) * Drawing(s) No(s)

   ....

   ....

   (8) Other documents listed below.

3. We agree that the proper law of this Contract shall be English Law unless the Works are situated in Scotland when the Contract shall in all aspects be construed and operate as a Scottish contract and shall be interpreted in accordance with Scots Law.

4. We have obeyed the rules about confidentiality of tenders and will continue to do so as long as they apply.

Formation of the contract 25

5. We undertake:-
(a) to satisfy the Secretary of State that the prices in the copy of the original Bills of Quantities which we will furnish within 4 days of request are those on which our tender is based and that they bear reasonable relation to each other;
(b) to provide a Programme within 4 days of request;
(c) to complete the Works to the satisfaction of the PM within the time stated in the Abstract of Particulars.

6. We agree that, should obvious errors in pricing or errors in arithmetic be discovered in any priced Bills of Quantities or Schedules of Rates submitted by us during consideration of this offer, we will be afforded the opportunity of confirming our offer, or of amending it to correct such errors.

7. Subject to and in accordance with paragraphs 3, 4, 5 and 6 above and the terms and conditions contained or referred to in the documents listed in paragraph 1, we offer to execute all the Works referred to in the said documents in consideration of:-

   (a) payment by the Authority of the sum shown in the Bills of Quantities of (in words)
   ....

   (b) reimbursement by the Authority of Value Added Tax in accordance with supplementary Condition No 139A.

   ALTERNATIVE 1

   Signed .... Secretary/Director ....

   Full name in capitals ....

   ALTERNATIVE 2

   The Corporate Seal of .... Ltd was hereunto affixed in the presence of:

   Signed .... Secretary/Director ....

   Full name in capitals ....

   Telex .... Postal Address ....

   Facsimile No ....

   Telephone No ....

   ....

   Date ....

Fig. 2.1: Form of Tender for use with GC/Works/1 conditions (firm BQs) — Form C1009T

contractors on Form C1009T which is reproduced as Figure 2.1. The Agency's invitation to tender fully describes the works and includes specification, drawings and bills of quantities together with the abstract of particulars.

As will be seen from Fig. 2.1, the tenderer offers to carry out and complete the works referred to in the tendering documents to the satisfaction of the Project Manager (PM) within the time stated for the sum shown in the bills of quantities, i.e., his tender price. The contractor also undertakes to satisfy the Authority that the prices in the bills of quantities which he will furnish to the Authority within four days of its request are those on which the tender is based and that they bear reasonable relation to each other. He further undertakes to provide a programme within four days of being requested to do so.

There is provision in GC/Works/1 (clause 33) for the programme to be submitted within 21 days after acceptance of the tender, but it is believed that in the vast majority of cases the programme will be required for submission with the tender or shortly thereafter.

The other provisions in the form of tender as illustrated are self-explanatory.

The Authority's acceptance of the contractor's tender brings a binding contract into existence, as echoed in the clause 1(1) definition of 'the Contract' already quoted. It is not, therefore, necessary for articles of agreement incorporating the conditions to be formally executed by the parties since contract conditions may be validly incorporated by reference, see e.g., *Modern Buildings (Wales) Ltd* v. *Limmer & Trinidad Co Ltd* (1975), and this is done by Form C1009T (see Fig. 2.1).

In fact it is desirable that formal articles of agreement incorporating GC/Works/1 should be prepared and executed, together with an appropriate attestation clause. This should certainly be done if GC/Works/1 is used in the private sector.

Contract documentation

From clause 1(1) we learn that the contract documents comprise:

- the contractor's tender and the employer's written acceptance
- the contract drawings
- a specification
- bills of quantities
- the printed conditions
- an abstract of particulars.

It is apparent from paragraph 2(8) of the form of tender (Fig. 2.1) that other documents as listed therein may also form part of the contract.

The abstract of particulars (Form C1009 Abs) is a vital and significant document and is included with the invitation to tender: clause 1(1). It fulfils some of the functions of the appendix in JCT 80 and related forms, and contains important terms and details. It identifies the Authority, the PM, the quantity surveyor and the person who will nominate the adjudicator, as well as giving dates for commencement and completion (see clause 34(1)), amount of liquidated damages etc. It can be used to modify the printed conditions. In the case of PSA projects, since the Agency has no separate legal personality, the 'Authority' is to be identified formally as the Secretary of State for the Environment, who is the minister in charge of the Agency.

The abstract of particulars will also reserve to the Authority any powers which the PM is not to exercise and which are reserved to the Authority, e.g. payment and retention, deduction of liquidated damages, and deciding on objections to nomination.

The addendum to the abstract of particulars (Form C1009 Abs-Add) will set out other important matters such as the time for nominations which is relevant in the case of a claim for prolongation and disruption under clause 46.

The Limitation Act 1980

GC/Works/1 can be entered into as a simple contract or as a specialty contract, i.e., a contract under seal or deed, although as already noted no formal provision is made for its execution in any particular way. It will, however, be seen from the form of tender itself (Fig. 2.1) that these alternatives are envisaged by the tendering documents. In fact, in modern practice it is not necessary actually to affix a seal, whether of wax or wafer, where a contract is entered into under seal (*First National Securities Ltd* v. *Jones* (1975)). It has also been held recently that failure to observe the exact formalities of sealing does not vitiate the intended making of a contract under seal. This was the holding in *Whittall Builders Co Ltd* v. *Chester-le-Street District Council* (1988), where an inadvertent omission to put an impression of a limited company's seal on a wafer did not prevent a building contract becoming and being the deed that was intended.

The making of a contract under seal has important practical consequences in regard to the Limitation Act 1980. It is the writer's firm view that *all* building contracts ought to be entered into under seal in order to protect the employer's legitimate interests.

The Limitation Act 1980, as amended by the Latent Damage Act 1986, prescribes the period within which actions on the contract must be commenced. It thus imposes a time limit within which an action for a breach of contract must be brought.

Under section 5 of the Act, 'an action founded on simple contract shall not be brought after the expiration of six years from the date on which the cause of action accrued', i.e., from the date when the breach of contract occurs.

Under GC/Works/1, it is submitted that, because of the contractor's obligation to carry out and complete the works (see clauses 34(1) and 7), in the case of defective work there are separate breaches both at the time of doing the work defectively and on completion. Consequently, time does not begin to run against the Authority until at least the date of completion as certified by the PM under clause 39(1).

Clause 21(1) deals with the contractor's liability to remedy defects which appear during the maintenance period(s) specified in the abstract of particulars. In respect of such defects only, if notified to him, the limitation period will run from when he is called upon by the Authority to make good the defects or at any rate from within a short time thereafter. If the defects remain undetected then the limitation period would run from the date of completion, if not from the end of the longest maintenance period when the PM is to issue a further certificate that the works are in a satisfactory state: clause 39(1).

Where the contract is made under seal, the limitation period is increased to 12 years. The limitation period again runs from 'the date on which the cause of action accrued': 1980 Act, section 8. The extended period of limitation is in itself a valuable protection to the employer since it appears that there is no longer dual liability on the contractor in contract and in tort: see, e.g., *Greater Nottingham Co-operative Society Ltd* v. *Cementation Piling & Foundations Ltd* (1988). The employer's remedies against the contractor (and *vice versa*) for breaches of contract lie only in contract lie only in contract even if the breach also amounts to negligence.

However, there are two situations in which the commencement of the limitation period may be extended. Both of them are relevant to building contracts generally and to GC/Works/1 in particular.

- *Postponement in case of fraud, concealment or mistake*

Section 32 of the Limitation Act 1980 provides that time shall not start to run where the right of action has been concealed by:

(a) the fraud of the defendant;
(b) the deliberate concealment by the defendant of any fact relevant to the plaintiff's right of action; or
(c) the action is for relief from the consequences of a mistake.

In that case, the limitation period does not start to run until the plaintiff has discovered the fraud, concealment or mistake or could, with reasonable diligence, have discovered it.

Both fraud and concealment are highly relevant to building and civil engineering projects and the operation of the section is well illustrated by *Applegate* v. *Moss* (1971). In that case, by a contract made in February 1957, the defendant agreed to build two houses for the plaintiffs and to support them on raft foundations reinforced with specified steel network. The defendant sub-contracted this work. The houses were completed towards the end of 1957 and the plaintiffs went into occupation. In 1965 serious cracks appeared in the buildings and it was discovered that the houses were in an unhabitable state because of the defective foundations which had neither raft nor reinforcing. The plaintiffs sued for breach of contract. Although their action was brought more than six years after the breach of contract they were successful, the Court of Appeal holding that there had been deliberate concealment by the builder within the meaning of what is now section 32. As Lord Denning MR so graphically put it 'the builder put in rubbishy foundations and then covered them up'.

Potentially, therefore, section 32 is of wide application in the construction industry since much defective work is covered up deliberately within the meaning of the clause. The plaintiff must establish a *prima facie* case of deliberate concealment and demonstrate that there is a triable issue in relation to it if he is met with a plea by the defendant that the action is statute-barred: *E. Clarke & Sons (Coaches) Ltd* v. *Axtell Yates Hallett* (1989).

- ***Action on an indemnity***

Clause 19 of GC/Works/1 is an indemnity clause since under it the contractor undertakes, *inter alia*, to reimburse the Authority in respect of any costs or expenses which the Authority incurs in dealing with or in settling any third party claim or proceedings in respect of any loss or damage which arises out of or is connected with the execution of the contract.

In *County & District Properties Ltd* v. *C. Jenner & Son Ltd* (1976), which concerned the indemnity clause of the former NFBTE/FASS form of nominated sub-contract, it was held the cause of action on the indemnity clause did not arise until the liability of the main contractor (who was seeking to enforce the indemnity against the sub-contractor) had been determined. It is accordingly the case under GC/Works/1, clause 19, that the Authority has six years from the date when the right to be indemnified arises in which to issue a writ and this can, in appropriate cases, postpone the commencement of the limitation period.

Proposals for reform

It is generally considered in the construction industry that the law relating to limitation of actions is defective and in late autumn of 1989 the Department of Trade and Industry published a report into the nature, scope, extent and implications of professional liability based on an inquiry by three study teams: *Professional Liability* (1989, HMSO, London).

The recommendations of the Construction Professions Study Team support the proposal that consideration be given to altering the law by amending the Limitation Act 1980 to provide:

- A limitation period for negligence actions in tort and for actions for breach of contract (whether or not under seal) of 10 years from practical completion or effective occupation.
- The 10 year period should be a longstop extinguishing the right, i.e., negative prescription as in Scottish law.
- 'Deliberate concealment' to be redefined so that ordinary construction processes would not lead to an exception to the longstop.

Whether or not these proposals will be put into law is a matter of some doubt in view of the negative response of the Government to the Report. However, the solutions suggested by the Report are more rational for the construction industry than those presently on the statute book.

Chapter 3

Contract documentation, information and staff

Introduction

This first section of the form consists of six clauses, the majority of which derive from earlier editions. GC/Works/1 envisages the use of exhaustive contract documentation consisting of (clause 1(1)) the set of general conditions, the abstract of particulars, a separate specification, drawings, bills of quantities and, of course, the contractor's tender and the Authority's written acceptance.

Many of the problems which arise under GC/Works/1 or any building or engineering contract result from the meaning to be given to the words used in the contract documentation. Problems frequently arise in the case of contract documents which are drafted by non-lawyers.

For example, in *Convent Hospital Ltd* v. *Eberlin & Partners* (1988), the court was faced with difficulties in interpreting a provision drafted by quantity surveyors in a bill of quantities which required the contractor to provide for a performance bond. The wording of the bill item and a note to it were both ambiguous. Judge James Fox-Andrews QC, Official Referee, was highly critical of the draftsmanship, and in particular of wording which required 'an appropriate deduction' to be made at final account stage if the bond was not in fact required. 'Appropriate reduction' might mean the contractor's actual price against the item or it could mean a greater deduction from the contract sum if the contractor had underpriced the item. Such ambiguities are common in bill provisions.

The basic rule of interpretation is that the contract documents must be read and construed as a whole, but this is subject to clause 2(1) which provides that, in case of conflict between the printed conditions and other contract documents, the printed conditions are to prevail. This seems to be an unsound policy which can defeat the true intentions of the parties: see, for example, *English Industrial Estates Corporation* v. *George Wimpey & Co. Ltd* (1972) on a similar provision in a JCT contract. The normal rules of legal interpretation will usually accord a higher priority to documents such as the tender and acceptance, if incorporated, or to other documents specially prepared for the individual project like bills or specifications. Clause 2 also contains further rules as to the precedence of documents, as discussed below. Clause 1(2) says that the headings to the conditions are not to affect their interpretation.

The *Eberlin* case is also of interest in the present context since it considered a method commonly used by contractors in pricing bills when preparing a tender. As the judge pointed out, by the priced bills of quantities the contractor is representing that he will undertake to carry out the various items at the prices shown, and one of the purposes of the priced bill is to enable the employer's quantity surveyor to determine whether it is likely that the contractor has put in an economically viable tender.

The learned judge said:

'These contractors operated on the basis that in pricing the bills in these circumstances their quantity surveyor sought to identify those items which might be omitted and putting a lesser price than that of the internal pricing therefor. The difference between the net cost to the contractor of such an item and the price put in the priced bill would then be transferred to and built into other rates. Ideally from the contractors' point of view enhanced rates would be entered for items where it was likely during the course of the contract the quantities required would be increased . . . . on the evidence before me I do not consider that this was improper conduct on the part of the contractors. The representations they were making were in no way false. They took the risk that financial loss would result to them if an item was priced below cost and the item was not omitted. Part of the skill required of the contractors was that the enhanced rates for other items should not be too easily identified and challenged.'

Under PSA tendering procedure, of course, the contractor's tender contains an undertaking (paragraph 5(a) of Fig. 2.1) that the prices in the bills on which the tender is based 'bear reasonable relation to each other'. However, this does not appear to militate against the common practice in pricing bills which has received judicial approval.

Clause 1 Definitions etc.

The definitions in clause 1 must be borne in mind in intrepreting the conditions which follow. Many of the definitions are of considerable significance and differ from those in clause 1 of Edition 2. Not all of the definitions are particularly helpful but they are applicable wherever the word or phrase in question appears in the contract since, unlike Edition 2, they are not prefaced by the words 'unless the context otherwise requires'.

Many of the definitions are self-explanatory and need little comment, but readers should note the following:

(i)
'The Abstract of Particulars' can be used to modify the conditions. It will identify the Authority, the Project Manager (PM), and the quantity surveyor (QS), and may also contain restrictions or limitations on the extent of the PM's authority as agent of the Authority: see clause 4(1). The abstract will also specify the period or periods for possession (see clause 34(1)) and the date of completion, the amount of liquidated damages (see clause 55(1)) and the length of the maintenance period(s) (see clauses 21(1) and 49(5)) and any sections of the works for which there is a separate date for completion. It will also name the person who is to appoint the Adjudicator (clause 59(1)) and deal with insurance (clause 8).

Although the conditions are silent on the matter, it is submitted that if the named PM or QS ceases to act by reason of death or incapacity or some other like reason, it is an implied term of the contract that the Authority will name a successor forthwith. Its failure to do so would constitute a breach of contract on its part: *Croudace Ltd* v. *London Borough of Lambeth* (1986) CA.

(ii)
The definition of *'the Accepted Risks'* is the same as that in Edition 2 after Amendment No. 4 of 17 November 1987. The definition is important when considering the contractor's liability under clause 19 for loss or damage arising out of or connected with the execution or purported execution of the contract, his obligation to insure under clause 8(1), and his duties under clause 13 (protection of works).

'The Accepted Risks' are those risks which may affect the works but which are outside the control of either of the contracting parties and for which the Authority accepts responsibility. Reference should be made to clause 19(5)(b) under which the Authority reimburses the contractor for any loss or damage so caused. The contractor is also entitled to an extension of time under clause 37(2)(d) for any delay caused by the occurrence of an accepted risk.

The list of 'accepted risks' is similar to those in other modern standard forms of construction contract.

(iii)

'*The Authority*' is the term used in the contract to mean the employer and he must be formally identified in the abstract of particulars. 'The Authority' must have a separate legal personality, e.g., the appropriate minister of the Crown. Many important functions are reserved to the Authority by the conditions. Under the terms of clause 4(1) the Authority may delegate its powers to make any decision to a person authorised by it. It is important that any delegations be notified in writing to the contractor as soon as possible, together with any qualifications or limitations imposed on the delegate's authority.

(iv)

The definition of '*the Contract*' also defines and governs the contract documentation. In addition to the seven documents in the definition, any document listed in the contractor's form of tender (see paragraph 2(8) thereof — Fig. 2.1) will also form part of the contract. Clause 2(1) gives the printed conditions priority in interpreting the contract over all other contract documents, which is an extension over the more limited precedence given to the printed conditions by clause 4 of Edition 2.

(v)

The definition of '*the Contract Sum*' is important, but is in very limited terms. It contrasts with the more extensive definition in Edition 2, clause 1(2), and also with the wider definition given in the first consultative draft of Edition 3 which was:

'the sum agreed by the Authority, or the sum calculated in accordance with the Contract and payable to the Contractor for the full and entire execution and completion of the Works before taking into account the effect of'

specified conditions adjusting the contract sum. The definition must be read in conjunction with that of '**the Final Sum**' which is defined as the 'amount payable under the Contract by the Authority to the Contractor *for the full and entire execution and completion of the Works*', i.e., after all necessary adjustments have been made under those provisions providing for adjustment or alteration of the contract sum.

(vi)
The definition of '*the Contractor*' extends to include 'legal personal representatives', i.e., a deceased person's estate's representatives on death. This is only applicable in the unusual case where the contractor is an individual rather than a limited liability company. The term also extends to 'permitted assigns', which means those assignees to whom the Authority has given consent under clause 61. If the contractor were an individual it is clear that his legal personal representatives would be entitled as of right to continue with the contract since a construction contract is not one calling for personal performance as is, for example, a contract to write a book.

(vii)
The definition of '*the Date or Dates for Completion*' is by reference to the dates set out or ascertained in accordance with the abstract of particulars, as adjusted by any extensions of time granted under clause 36, is necassary because of the provisions of clauses 34 (commencement and completion), 36 (extensions of time), 38 (acceleration), 52 (cost savings), 55 (liquidated damages), and 56 (determination). The contractor's obligation is to complete the works 'to the satisfaction of the PM by the Date or Dates for Completion': clause 34(1).

(viii)
'*The Project Manager*' must be named in the abstract of particulars. He is the person charged with 'managing and superintending the Works', i.e., he exercises the day to day role of the Authority and superintendence on its behalf. Any limitations or restrictions on his authority must be set out in the abstract of particulars since otherwise (clause 4(1)) he is 'deemed authorised to act generally for the Authority' which, without more, would mean that he could make any decisions which are to be made by the Authority under the contract. He has extensive express powers under the contract: see, in particular, clause 40 (PM's instructions). In some cases, e.g., as to whether any instruction is necessary or expedient (clause 40(3)) his decisions are made 'final and conclusive' and so cannot be reviewed in arbitration.

(ix)
'The Programme' is defined in general terms as meaning 'the document or documents submitted prior to the acceptance of the tender and agreed at that time by the Authority or agreed by the Contractor and the PM' under clause 33. This sets out the detailed provisions for the programme. Read together, the definition and clause 33 require the programme to be a document of considerable depth and sophistication. It is assumed that in the majority of cases it will be required to be submitted with the tender and its form will be specified in the Authority's invitation to tender. The importance of the programme (see the discussion on clause 33) means that a simple bar chart will not be sufficient since the programme must be fully resourced and show, *inter alia*, the intended sequence of the works. Network analysis or critical path programmes will probably be required.

(x)
The definition of *'the Site'* is based on that in clause 1(2) of Edition 2. The definition is important when read in conjunction with clause 34(1) since both the contractor's obligation to commence the works and the date of completion are dependent upon the Authority granting the contractor possession of the site. It is essential that the site be clearly identified and its boundaries delineated in the abstract of particulars and that, in appropriate cases, the abstract should address the problem of a small part of the site not being available on the date for possession when the major part was. The definition may be contrasted with that in clause 1(1)(n) in the ICE Conditions, 5th Edition, where 'the site' means 'the lands or other places on under in or through which the Works are to be executed and any other lands or places provided by the Employer for the purposes of the Contract'.

(xi)
'The Stage Payment Chart' as here defined is important since advances on account, i.e., interim payments, are to be paid in accordance with the stage payment chart or charts which must be included with the invitation to tender. At the time of writing, the PSA has not finalised the form of the stage payment chart but it is probable that such charts will be expressed in percentage terms of the contract sum. The charts will be based on a pre-determined 'S' curve plotted following a study of more than 200 PSA projects.

(xii)
The definition of *'the Works'* is substantially the same as that in Edition

2 and emphasises that 'the Works' are not only those described in the specification and/or bills and/or shown on the drawings, but also include 'all modified or additional works to be executed under this Contract', i.e., as a result of a variation instruction issued by the PM under clause 40(5).

Clause 1 (2) is a usual type of interpretative provision and says that the headings to the conditions are not to be considered in interpreting the conditions.

There are numerous important notices which must or may be given under the contract including those from the contractor requesting an extension of time (clause 36(1)) or notifying a claim for prolongation or disruption under clause 46(3), or from the Authority determining the contract under clause 56(1).

Clause 1(3) makes clear that any such notice is to be in writing, which means that the words must be in visible form — handwriting, typescript, printing, telex, facsimile transmission etc. — as contrasted with oral communication. A notice may be given to the contractor or his agent (see clause 5) or sent by post to the contractor's registered office or last known place of business. If it so posted it is 'deemed to have been served on the date when in the ordinary course of post it would have been delivered'. This means the next Post Office working delivery day after posting and the contractual presumption may in fact have no basis in reality, having regard to the continual worsening of the postal services in the United Kingdom.

Clause 1(4) is extremely important in view of the very strict time limits imposed under the contract. For example, clause 46(3)(b) imposes a time limit of 56 days for the contractor to provide full details of all expenses incurred as a consequence of disruptive events and an even stricter time limit is imposed on the quantity surveyor (clause 46(5)) who must notify the contractor of his decision on the contractor's claim for prolongation or disruption within 28 days of receipt of this information. In practice, these and other time limits may be difficult to meet, and consequently this provision enables the parties to agree that the time periods be extended, even if the specified period of time has expired. This cannot, however, be done in relation to the periods specified in clause 36 (extensions of time) which are sacrosanct and cannot be waived, by agreement or otherwise. It is suggested that any agreement under clause 1(4) should be in writing.

There is no definition of the word 'day', and this is presumably a deliberate omission. The word, therefore, has its usual meaning and not a working day, and this is of significance when one considers the many strict time limits imposed by the Contract.

Clause 2 Contract documents

This important clause deals with priority within the contract documents (see clause 1(1) for the definition of 'the Contract') and of discrepancies in and between them, as well as with the documents which are to be made available to the contractor. The contract proceeds on the assumption that the contract documents should be consistent, but in the nature of things inconsistencies in or between the contract documents are not uncommon.

Clause 2(1) is more wide-ranging than was clause 4(1) of Edition 2 and the extension is much to be regretted. It now provides that in case of discrepancy between the printed conditions *and any other documents forming part of the contract*, the provisions of the printed conditions are to prevail. The normal and sensible rule of contractual interpretation is that specially prepared documents such as bills of quantities take precedence over printed conditions (see *Love & Stewart Ltd* v. *Rowtor SS Co. Ltd* (1916)) which is presumably what the parties intended. The effect of clause 2(1) is to displace this sensible rule and make the printed conditions prevail over any specially prepared documents.

However, clause 2(1) does not prevent additional or supplementary terms being written into the other contract documents, provided these do not conflict with any of the printed conditions: see *Glenlion Construction Co. Ltd* v. *The Guinness Trust* (1987). However, it does mean that if there is any conflicting term — for example in the bills of quantities — or any qualification or amendment to the conditions in the other contract documents, the printed conditions will apply. If it is desired to amend the printed conditions this should be done by actually deleting and/or amending the printed text itself, and initialling the alteration.

The courts have considered the effect of the equally unmeritorious and even more widely drawn provision in the JCT standard contracts in numerous cases and have held that such words must be given effect. For example, in *Gold* v. *Patman & Fotheringham Ltd* (1958), the somewhat similar JCT provision was held to nullify a clause in the bills of quantities which was in conflict with the printed conditions. Lord Hodson said:

'The paragraph [in the bills] beginning with the words "such insurance" purports to override the provisions as to insurance contained in condition 15 and must therefore be disregarded, having regard to condition 10.'

In *English Industrial Estates Corporation* v. *George Wimpey & Co. Ltd* (1973), whilst giving effect to the similar JCT clause, the Court of Appeal was not happy about it, and while Lord Denning MR took the view that specially prepared provisions should always prevail over printed conditions, the majority held that the specially prepared terms must give way to the general printed conditions. The effect is that none of the other contract documents can be used to modify the printed conditions, except that 'they may add obligations which are consistent with the obligations imposed by the conditions': *per* Lord Justice Stephenson in *English Industrial Estates Corporation* v. *George Wimpey & Co. Ltd* (1973).

Clause 2(2) says that the specification is to take precedence over the drawings unless the PM instructs otherwise. This is a new and sensible provision, the effect of which is that the detailed words of the specification will take precedence over any drawings or notes on them.

Clause 2(3) deals with the contractor's obligation with regard to discrepancies between the specification and drawings or between different parts of the drawings and different parts of the specification. The contractor is bound to inform the PM of any discrepancies which he discovers when handling the documents in order to prepare or carry out the works. However, it is clear that there is no positive duty on him to search for discrepancies or minutely to examine the documents and accept the financial consequences if he fails to spot a discrepancy: see *London Borough of Merton* v. *Stanley Hugh Leach Ltd* (1985). If the contractor notifies the PM of a discrepancy, then the PM must issue an instruction under clause 40(1) to deal with it.

Clause 2(4) is in terms identical to clause 4(2) of Edition 2 and says that figured dimensions on drawings are to be followed in preference to the scale.

By *Clause 2(5)* the PM is to provide the contractor free of charge with a copy of each of the contract drawings, the specification, the unpriced bills, and any drawings issued during the progress of the works. The contractor must keep one copy of *all* drawings and of the specification on site. The PM and his representatives are to have access to these documents 'at all reasonable times', i.e., during normal working hours.

It is clearly an implied term of the contract that the PM will provide the contractor with correct information and drawings as the work proceeds, as part of the more general term implied on the part of the Authority that it will not hinder or prevent the contractor from carrying out his obligations in accordance with the terms of the contract and from executing the work in a regular and orderly manner. The position was well put by Mr Justice Vinelott in *London Borough of Merton* v. *Stanley Hugh Leach Ltd* (1985) when, dealing with the position under a JCT 63 contract, he said:

'The implementation of a building contract embodying the JCT conditions does require close co-operation between the contractor and the architect ... The implied undertaking by the building owner extends to those things which the architect must do to enable the contractor to carry out the work and the building owner is liable for any breach of this duty on the part of the architect.'

This applies with equal force to a contract on GC/Works/1 conditions, making the necessary substitution in the quotation of the words 'the Authority' for 'building owner', and 'PM' for 'architect'.

Clause 3 Bills of quantities

The published version of Edition 3 is designed for use with bills of quantities where all or most of the quantities are firm and not subject to re-measurement. The principal characteristic of a lump sum contract based on firm bills is that the contractor contracts, for the contract sum, to carry out only the work in the bills and any adjustment that involves a change from what is set out in the bills will involve an adjustment of the contract sum. This has been subject to some criticism by some commentators, and notably by the learned editor of *Hudson's Building Contracts*, 10th edition (1970, Sweet & Maxwell, London), chapters 5 and 8.

The disadvantages of the bills of quantities system can be very much overstated and it is submitted that they are greatly outweighed by the considerable advantages to both parties of removing a substantial area of possible dispute. The purpose of clause 3 is to define the effect of the bills of quantities and this it does in terms which are not dissimilar (though they are much better drafted) than the comparable provisions in JCT 80, clause 2.2.

Clause 3(1) is identical in effect to clause 5(1) of Edition 2. By its terms, the bills of quantities are *deemed* to have been prepared in accordance with the method of measurement expressed in them which, in the case of building works, will normally be SMM7 and in the case of civil engineering work will be CESMM (2nd edition, 1985). It is submitted that as a matter of good practice any departure from the method of measurement used in the bills should be clearly identified and explained.

Clause 3(2) deals with errors in description or in quantity in the bills and any omissions in them. No such omissions or errors shall vitiate, i.e., invalidate, the contract or release the contractor from any of his contractual obligations to execute and complete the works or release him from any of his other contractual obligations or liabilities.

By *clause 3 (3)* errors or omissions in the bills of quantities are to be rectified by the Authority under clause 42(4), i.e., they are to be treated as a variation and valued by the quantity surveyor in accordance with the rules laid down in that sub-clause. The contract sum is to be adjusted accordingly.

The proviso to this sub-clause is of vital importance. Errors, omissions or wrong estimates in the contractor's bill prices or in his bill computations or calculations are not to be rectified and the contractor is bound by them. This, in any event, is the position under the general la   since if a contractor makes a unilateral mistake in his tender price o₁ individual errors in pricing, multiplication, or addition, he is bound by them unless the employer or his agent discovers the error before acceptance and realizes that it is not intentional. *W.Higgins Ltd* v. *Northampton Corporation* (1927) is decisive on that question in the case of a lump sum contract such as GC/Works/1.

In that case Higgins contracted with the Corporation for the erection of 58 dwellings and completed the tender incorrectly. Higgins thought that he was tendering to build the houses at £1670 a pair when in fact he had offered to do so for a price of £1613 because of the way in which he had priced the bills. Mr Justice Romer accepted that Higgins' mistake was bona fide and that 'the result was very unfortunate for the plaintiff and an extreme hardship on him'. He nonetheless held that Higgins was bound by his error and refused to order rectification of the contract; see also *Riverlate Properties Ltd* v. *Paul* (1974), CA.

The position is different, it seems, if the employer discovers an error in the contractor's tender before acceptance and the argument that, in that case, the employer may keep the contractor to the mistaken price is unsound. Thus, in the Canadian case of *McMaster University* v. *Wilchar Construction Ltd* (1971), the employer purported to accept a tender knowing that the contractor had omitted the entire first page of his bid which contained an intended price variation clause. The High Court of Ontario held that the contractor was not liable. The contractor's mistake was as to the terms of the offer itself and not merely as to the motive or underlying assumptions on which the offer was based. As Mr Justice Thompson expressed the point:

> 'This is a case where one party intended to make a contract on one set of terms and the other intended to make it on another set of terms. The parties were not *ad idem*. The existing circumstances prevented the formation of a contract.'

The cases of *Collen* v. *Dublin County Council* (1908) and *Neill* v. *Midland Railway Company* (1869) are to the same effect. Whether an erroneous bill rate applies to additional work will depend on the construction of the bill: see *Dudley Corporation* v. *Parsons & Morrin Ltd* (1959).

Clause 3(4) covers the situation where there are approximate quantities in the bill. They do not gauge or limit the amount or description of work to be executed by the contractor, but in the final account they are to be replaced by actual quantities measured by the quantity surveyor and priced at bill rates and prices. The final sentence permits the quantity surveyor to move away from the bill rates and prices where he 'is of the opinion that because of *substantial* difference between billed and measured quantities it would be *unreasonable* to use the rates and prices in the Bills'. In that event, he is to value by measurement and valuation at rates and prices deduced or extrapolated from the bill rates and prices (clause 42(4)(b)) or at fair rates and prices having regard to current market prices (clause 42(4)(c)). The use of the word 'extrapolated' will, it is submitted, enable the quantity surveyor to deduce a rate which is outside the range of rates for similar work in the bills. The verb 'to extrapolate' means, in mathematical terms, to estimate the values already known by the extension of a curve or 'to infer [something not known] by using but not strictly deducing from the known facts': Collins' *English Dictionary*, 2nd edition (1986, Collins, London). It seems that the quantity surveyor will only move into the fair rates situation where it is not possible to deduce or extrapolate.

Clause 4 Delegations and representatives

This important revised condition deals expressly with delegations to and by those acting on behalf of the Authority. It delimits the actual and ostensible authority of the PM and other representatives of the Authority. It is the 'ostensible or apparent authority of the agent as it *appears to others*', as Lord Denning observed in *Hely-Hutchinson* v. *Brayhead* (1968) which is important.

Clause 4(1) establishes that the PM is the general agent or representative of the Authority and is deemed authorised 'to act generally for the Authority *subject to any exclusions set out in the Abstract of Particulars*'. If there are no exclusions in the abstract then the PM can make any decision which is to be made by the Authority under the contract. The first sentence enables the Authority, i.e, the employer, to delegate its decision-making powers to any person such as an authorised officer of the PSA.

In practice it is thought that the PM will not take on all the powers of the Authority under the contract and the abstract of particulars will reserve various matters to the Authority acting through its authorised representatives. For example, it is to be expected that there will be reserved to the Authority such matters as disputes about certificates (clause 39(2)), set-off (clause 51), determination (clause 56(1)) and matters arising under the arbitration agreement (clause 60).

Both the PM and QS (see clause 1(1)) are empowered by *clause 4(2)* to delegate in writing any of their duties and powers, and it is envisaged that a clerk of works or resident engineer may be appointed. If so appointed, the clerk of works or resident engineer is to execute the PM's powers under clause 31 (quality) which relate to the quality of work and materials, but they are not necessarily so limited since they may exercise 'such other powers as the PM may delegate to' either of them.

Where the Authority appoints a clerk of works or, it is submitted, a resident engineer, it is vicariously responsible for any negligence by him in the performance of his duties: *Kensington & Chelsea & Westminster Health Authority* v. *Wettern Composites and Ors* (1984). The contractor should ensure that all such delegations are, in fact, given in writing as is required by *clause 4(4)* which says that the contractor must be notified of all delegations, and any subsequent changes, as soon as possible. Such notification must be in writing: clause 1(3).

It is to be noted that, as a result of *clause 4(3)*, the Authority, PM or QS can exercise directly such delegated powers or duties, e.g., if the delegates fail to exercise their authority properly or at all, or if it is desired to vary any instruction given by the clerk of works or resident engineer.

The quantity surveyor has many important and separately identified functions under the contract, in particular those relating to valuation (clauses 41 to 43), prolongation and disruption claims (clause 46) and the final account (clause 49). However, it is clear that the quantity surveyor is part of the Project Manager's team and, while he is expected to act independently in the exercise of his duties, it is thought that the PM is not bound by a valuation made by the quantity surveyor and he must in fact be satisfied that the valuation was properly made. Reference should be made to *R.B.Burden Ltd* v. *Swansea Corporation* (1957), HL, where the analogous case of the architect and quantity surveyor under a JCT contract is discussed. It is submitted that the views of Lord Radcliffe, as expressed in that case, are of equal application to the relative positions of the PM and the quantity surveyor under a GC/Works/1 contract. His lordship said that there is nothing 'in the contract which suggests that the architect is bound to accept the quantity surveyor's opinion or valuation when he carries out his own function of certifying sums for payment' and while advances on account (clause 48) are not dependent on a certificate of the PM, they are dependent on 'any relevant instructions' being complied with and on 'all the work to which the advance relates' being to the satisfaction of the PM.

Clause 5 Contractor's agent

This clause derives from clause 33 of Edition 2 and obliges the contractor to employ a *competent* agent to supervise the execution of the works. He is to be on site 'during all working hours' and, if required to do so, he must attend the PM's office. Any notice given to him is good delivery to the contractor (clause 1(3)). The contractor's agent is the contractor's permanent representative on site. Apart from supervising the execution of the works, the contractor's agent must attend regular progress meetings to assess the progress of the works (see clause 35) and thus has a key role to play in the successful completion of the project.

It is suggested that the contractor should inform the Project Manager of the name of the competent agent (if he is not one of the key personnel named in the tender (see clause 6)). If there is any change in his identity, the PM should be notified of the change. In practice, it will be the agent who will receive those limited instructions which may be given orally under clause 40(2)(b),(d),(g) and (k), e.g., execution of emergency work under clause 54. If such instructions are given verbally they must be confirmed in writing within seven days. If they are not so confirmed, the contractor should seek written confirmation from the PM.

Clause 6 Contractor's employees

Clause 36 of Edition 2 empowered the Authority and the then superintending officer to require the contractor to cease to employ any employee in connection with the contract and to replace that person with someone else. It drew a distinction between grades of employee, the replacement of some of whom could be required by the SO and others by the Authority.

Clause 6(1) does away with this unnecessary distinction by empowering the PM to require the contractor to cease to employ in connection with the contract 'any person . . . . whose continued employment is in [his] opinion undesirable' and to replace him with 'a suitably qualified person'.

The importance of this provision from a management point of view is emphasised by the fact that this is one of the occasions when the decision of the PM is made 'final and conclusive'. The finality of the PM's decision extends to both the question of the continued employment of the person being undesirable *and* to the suitability of any proposed replacement.

This power is not one to be exercised capriciously and without reason and is clearly aimed at such matters as patent incompetence, unsafe working practices and things of that sort.

Clause 6(2), which is entirely new, recognises the modern practice of naming or agreeing a team of personnel prior to the award of the contract. If any personnel are named in the tender, any changes to them are to be notified to the PM.

Chapter 4
General obligations

Introduction

The 19 clauses in this section of the form are disparate and are grouped under one heading for convenience. Some of them are *conditions* of the contract in the legal sense, i.e., major terms, the breach of which may give rise to a right to treat the contract as repudiated. Others are minor or subsidiary terms, ('warranties'), i.e., stipulations the breach of which may give rise to a claim for damages but not to a right to treat the contract as repudiated.

This simplistic approach to the classification of contract terms is taken from the law of sale of goods, but a distinction must always be drawn between major and minor terms and the analysis is good for present purposes: see *The Mihalis Angelos* (1971). For a comprehensive discussion of the relative importance of contractual terms see *Cheshire, Fifoot Law of Contract*, 11th edition (1986, Butterworths, London) pp.140–149.

Into the category of major terms must be placed clauses 7 (conditions affecting works), 10 (design) and 21 (defects). Many of the other provisions are self-evidently of a subsidiary nature. Many of them, too, are drawn from Edition 2 of GC/Works/1, but all of them impose obligations on the contractor.

Table 4.1 summarises the contractor's duties under GC/Works/1.

General obligations

Table 4.1: Contractor's duties under GC/Works/1 — Edition 3

Clause	Duty	Comments
2(3)	Draw the PM's notice to discrepancies in or between specification and drawings.	
2(5)	Keep one copy of drawings and specification on site.	PM or his representative to have access at all reasonable times.
5	Employ a competent agent to supervise the works.	He must attend site in working hours and attend the PM's office as required.
6(1)	Replace any employee if so required by the PM.	He must be replaced by a suitably qualified person.
6(2)	Notify the PM of any changes.	If required in his tender to supply names of personnel.
7(1)	Satisfy himself as to: • Roads, rail etc. • Contours and boundaries. • Risk of injury to adjacent persons or property. • Nature of soil. • Conditions for execution of work and precautions to prevent nuisance etc. • Supply of labour. • Supply of materials. • Anything else likely to affect the work or the price.	In relation to site.
7(3)	By notice immediately: • inform the PM; and • state the proposed remedial measures.	If contractor becomes aware of unforeseeable ground conditions.
8(1)	Effect and maintain: • Employer's liability insurance for employees. • Insurance against loss or damage to works and 'things' for which the contractor is responsible. • Insurance against personal injury and damage to property.	For duration of contract and longest maintenance period.
8(2)	Maintain insurance in respect of fire, damage to works etc. and injury to persons or property. Send to authority, certificate in form attached to abstract of particulars that policy has been effected	As notified in his tender. Within 21 days of acceptance of tender
8(3)	Maintain insurance in accordance with the summary of essential requirements attached to the abstract of particulars. Send to authority, certificate in form attached to abstract of particulars.	To jointly cover the authority, contractor and all sub-contractors. Within 21 days of acceptance of tender
8(4)	Give notice to authority that policies have been renewed.	If insurance is effected by annually renewable policy and works are not complete by renewal date.

Table 4.1: Contractor's duties under GC/Works/1 — Edition 3 (cont.)

Clause	Duty	Comments
	Produce evidence that alternative fully equivalent cover has been arranged.	If policies no longer exist or contractor knows they are ineffective
9(1)	Set out the works.	PM must provide all necessary information.
9(2)	Provide, fix and maintain stakes, level marks etc. Take precautions to prevent removal.	He is liable for the consequences of removal.
10(1)	Submit design drawings, etc to the PM for approval. Not to commence work until written approval is obtained. Not to alter the design without further written approval	If contractor is to design any part of the works.
11(1)	Give all notices. Pay all fees. Supply all drawings.	Required by Act of Parliament etc. Required by Act of Parliament etc.
11(2)	Obtain consent, permission or licence of statutory undertakers or adjoining owners. Pay any licence fee or charge.	Whose services or land is affected or whose consent is necessary.
12(1)	Pay royalty, licence fee or other expense.	
12(2)	Reimburse the authority for all claims etc.	In respect of royalties, fees etc
13(1)	Protect the works. Have custody of 'things.' Take reasonable steps to protect, secure, light, watch places about the site.	During the execution of works. Which may be dangerous.
13(2)	Comply with statutory regulations.	Governing storage and use of all 'things' brought on to site for the works.
14	Take precautions to prevent nuisance. Protect waterways from pollution.	
15	Provide the PM with return in form as directed of workpeople and plant employed each day.	
16	Not to lay foundations until excavations have been inspected.	
17	Give reasonable notice to PM when work or 'thing for incorporation' is to be covered.	In default, the contractor must uncover at his own expense.
18(1)	Send representative to take joint measurements for final account.	If required by the QS.
18(2)	Provide equipment for measuring the work.	Without extra charge.
19(2)	Reinstate, replace or make good any loss or damage arising from or in connection with the execution of the contract; *or* Compensate the Authority.	Without delay and at contractor's own cost. If the Authority agrees.

General obligations

Clause	Duty	Comments
19(3)	Reimburse authority for costs or expenses incurred in dealing with or settling proceedings.	If claim is brought against Authority in respect of loss or damage.
20(1)	Reimburse the authority against all claims in respect of damage or distress due to: • loss of personal data; or • disclosure of personal data in clause 20(2) circumstances.	
21(1)	Make good notified defects which appear during the maintenance period.	Without delay, at own cost and to the Authority's satisfaction.
22	Comply with rules of occupying body. Comply with changes in the rules.	If works executed in boundary of Government premises.
23(1)	Not to unlawfully discriminate.	Within the meaning of the Race Relations Act 1976 or statutory modification thereof.
23(2)	Take steps to ensure observance of 23(1) by employees, sub-contractors etc.	
24(1)	Not alone or with others to: • corruptly receive gifts; or • give gifts to a Crown employee, or consultant or contractor who has a contract to which the Crown is a party	As an inducement or reward etc.
25(1)	Keep necessary records.	For QS, PM or authority to verify claims.
25(2)	Afford the PM and QS access to clause 25(1) records and supply information required.	So that they can discharge their contract functions.
26(1)	Take steps to prevent unauthorised entry to site. Take steps to prevent a person from being admitted to site.	As required by PM. If PM so notifies the contractor.
26(3)	Give the PM a list of persons connected with the works specifying how so connected and other required particulars.	If so instructed by the PM.
26(4)	Bear cost of notice, instruction or decision of the PM under this clause.	
27	Submit list of names of workpeople to the PM and any other information reasonably required.	If passes required before admission.
28	Not to take photographs of the site or works. Take all steps to ensure that no such photographs are taken or published etc.	At any time without prior written permission of the PM
29(1)	Comply with the Official Secrets Act 1911 to 1939 and Section 11 of the Atomic Energy Act 1946.	Where appropriate.

Table 4.1: Contractor's duties under GC/Works/1 — Edition 3 (cont.)

Clause	Duty	Comments
29(2)	Use or disclosed contract information only in connection with the contract.	
30(4)	Forthwith remove unused or rejected 'things'.	At his own expense.
31(1)	Execute the works with diligence and expedition and in accordance with the programme. Exercise all proper skill and care in a workmanlike manner and to the satisfaction of the PM.	
31(2)	Ensure that: • the works • 'things' for incorporation are fit for purpose; *and* conform to contract documents.	Using skill and care of experienced and competent contractor
31(3)	Notify PM before incorporation of any 'things' the contractor thinks should not be incorporated.	
31(4)	Demonstrate to PM that contractor is performing his duties under clauses 31(1) and 31(2). Give PM assistance for inspection.	
31(5)	Bear costs of expert testing.	If 'thing' tested does not substantially conform to contract.
	Bear costs of further tests reasonably, required to monitor quality following negative test results.	Even if tests are satisfactory.
31(7)	Replace, rectify or reconstruct: • works which do not comply with the contract or the PM's satisfaction. • 'things' for incorporation' which do not comply with the contract or which have been rejected by the PM.	At his own cost.
32(3)	Take all practicable measures not to disturb the object. Cease work if danger to the object or likelihood of impeding removal. Take steps to preserve object in position and condition found. Inform PM of discovery and location.	Upon discovery of a fossil etc.
33(2)	Submit programme to the PM for agreement.	Not later than 21 days from acceptance of tender unless submitted before acceptance of tender and agreed by authority.
34(1)	Take possession of site. Commence the execution of the works. Proceed with diligence or as instructed by the PM.	After notice given. Forthwith. So that the works are completed to PM's satisfaction by date for completion.

General obligations

Clause	Duty	Comments
34(2)	Remove unused 'things for incorporation' and all 'things' not for incorporation. Keep site tidy and free from debris. Clear and remove rubbish. Deliver the site and works by the due date to the PM's satisfaction. Comply with any instructions relating to the removal of any things and rubbish.	Not later than the date for completion. At his own cost.
35(1)	Send agent to attend progress meetings.	To assess progress and assist completion by due date.
35(3)	Submit to the PM a report which: • describes progress in regard to programme and instructions • specifies contractor's outstanding requests for drawings, nominations, levels etc • explains delaying circumstances • refers to requested extensions of time • sets out proposals for programme amendments to ensure completion by due date.	Three days before each meeting.
36(5)	Submit claim to PM giving grounds for extension or further extension.	Not later than 14 days from receipt of decision if contractor dissatisfied. PM must notify contractor in 28 days whether decision amended.
36(7)	Endeavour to prevent or minimise delay. Do all reasonably required by PM to proceed with the works.	Contractor is not entitled to extension if delay due to his default, negligence, improper conduct or lack of endeavour.
40(3)	Comply with any instruction.	Forthwith.
40(4)	Not to add to, omit from, or otherwise alter the works except in accordance with an instruction.	
41(4)	Submit to the QS any information required to enable valuation of the VI.	
42(2)	Include with any quotation under clause 40(6) such other information as he considers will help the QS to evaluate the quotation.	
42(6)	Submit clause 41(4) information not later than 14 days after QS request.	
42(8)	Give reasons for disagreement and his own valuation within 14 days of QS notification.	If contractor disagrees with valuation, otherwise he is treated as having accepted it.
42(10)	Give reasonable notice to QS before commencing work and deliver vouchers in required form by end of week following week in which work was done.	If QS decides work will be valued in accordance with clause 42(4)(d).

Table 4.1: Contractor's duties under GC/Works/1 — Edition 3 (cont.)

Clause	Duty	Comments
43(2)	Submit clause 41(4) information within 28 days of compliance with instruction.	QS must notify contractor within 28 days of receipt, the amount determined.
43(3)	Notify QS of reasons for disagreement and submit his own estimate within 14 days of receipt.	If contractor disagrees with clause 43(2) amount, otherwise he will be regarded as having accepted the amount.
45(1)	Not to issue any claim inclusive of VAT. State how the work or supply is rated. Show relevant rates separately. Provide required further information.	When requesting payment.
45(2)	Satisfy himself that tax has been allowed for. Promptly issue a VAT receipt.	If so satisfied.
48(6)	Demonstrate to the PM that any amount due to a nominated sub-contractor or supplier of 'things for incorporation' covered by a previous advance has been paid.	If so requested before payment of an advance or issue of the final certificate.
49(2)	Notify agreement or disagreement with final account. Specify reasons.	Within three months of receipt. In any notice of disagreement.
49(5)	Pay the excess to the Authority.	If the PM has certified the works are in a satisfactory state and if the final sum has been agreed or treated as agreed and if the amount paid on an estimated basis exceeds the final sum.
52(2)	Provide further information relating to cost saving proposals.	As required by the PM.
52(3)	Continue the expeditious carrying out of the works.	Having submitted a proposal to the PM.
54(1)	Carry out emergency work.	If required by the PM.
54(2)	Reimburse costs incurred by the Authority in carrying out emergency work if they would have been avoided had the contractor carried out the work.	If the contractor fails to do the work.
55(4)	Pay the Authority the difference.	If liquidated damages exceeds any advance due.
56(5)	Promptly comply with Authority's directions under clause 56(3).	After determination by the Authority.

General obligations

Clause	Duty	Comments
57(1)(c)	Assign to the Authority the benefit of any sub-contract or supply contract.	After determination by the Authority unless for insolvency and without further payment.
57(3)	Pay the Authority the amount of the excess.	If the total cost of completion after determination exceeds the sum which would have been payable to the contractor.
58(4)	Refer the circumstances to the Authority.	If the contractor is of the opinion that his unavoidable losses and expense due to determination have not been fully reimbursed.
59(2)	Send copy of notice and enclosures to PM and QS at the same time.	If contractor gives notice to adjudicator.
61	Not to assign the contract.	Without the written consent of the Authority.
62(1)	Not to sub-let.	Unless Authority accepted sub-letting proposed before award of contract or specified in the contract or prior written consent of PM.
	Provide details of proposed sub-contractor.	As the PM requires.
62(2)	Ensure each sub-contract will enable him to fulfil his obligations under the contract. Require the sub-contractor to assume obligations owed to the Authority and PM by the contractor. Give sub-contractor rights against the contractor as the contractor's rights against the Authority.	To the appropriate extent.
62(3)	Take necessary action so that sub-contractor complies with his obligations.	Whenever the Authority requests and without prejudice to his obligations under other terms of the contract.
62(4)	Secure completion of sub-contract works at his own expense.	If sub-contract is determined due to default of sub-contractor.
63(2)	Not to order work or 'things' under prime cost items prior to the conclusion of a sub-contract.	Without the written consent of the PM.
63(3)	Produce to the QS quotations etc.	Necessary to show sums paid by the contractor in respect of prime cost items.

Table 4.1: Contractor's duties under GC/Works/1 — Edition 3 (cont.)

Clause	Duty	Comments
63(5)	Supply detailed schedule of rates properly used to calculate contract sum.	If bills of quantities, are not provided for work or 'things' to which this clause is applicable.
63(6)	Provide information requested by PM.	If contractor makes objection to person to be employed as a nominated sub-contractor.
63(7)	Not to determine or assign the sub-contract without the agreement of the Authority.	Once the contractor has entered into a contract with a nominated sub-contractor.
65(1)	Give reasonable facilities for other works executed by the Authority on the site.	

Clause 7 Conditions affecting works

Clause 7 covers the contractor's responsibility for anticipating conditions affecting execution of the works and for pricing accordingly. It is similar in terms to its predecessor, i.e., clause 2 of Edition 2, and the list of items is no different from those in the old provision. The revised condition includes provisions about unforeseeable ground conditions which were introduced into Edition 2 by an amendment in 1987.

The importance of this clause cannot be over-emphasised. It places on the contractor the risk that the site and allied conditions may turn out to be more onerous than he expected. It also makes plain that, except where he encounters 'unforeseeable ground conditions' as defined, he is not entitled to any additional payment on the ground of any misunderstanding or misinterpretation of the matters to which the clause refers, nor, as a result, is he to be released from any contractual risks or obligations or because he could not or did not foresee any matter affecting the works.

Clause 7 in effect merely restates the general and sensible rule of law. Under the general law, the employer does not warrant that the site is fit for the works or even that the contractor will be able to construct the building on the site to the design supplied (*Bottoms* v. *York Corporation* (1892); *Thorn* v. *London Corporation* (1876)), although in an appropriate case the contractor may have a claim against the employer for misrepresentations made about site and allied conditions during pre-contractual negotiations: see *Morrison-Knudsen International Co Ltd* v. *Commonwealth of Australia* (1972). The Authority could, therefore, be liable for misrepresentation both at common law and under the Misrepresentation Act 1967, as amended, despite the provisions of clause 7.

The case of *Bacal Construction (Midlands) Ltd* v. *Northampton Development Corporation* (1978) is of interest in this connection. There, tender documents stated that the site was a mixture of sand and clay, but in fact there were strata of tufa as well. At the time of tender, the contractors were informed in writing by the employer that their design for foundations was to assume soil conditions disclosed at boreholes. These gave no indication of the presence of tufa. The contractors were held entitled to recover compensation, the Court of Appeal holding that it was an implied term of the contract that the ground conditions would be in accordance with the information on which the contractors were instructed to design. In other words, the Authority may be taken to warrant the accuracy of any assumptions which it instructs the contractor to make for tendering purposes.

Clause 7(1) deems the contractor to have satisfied himself as to communication with and access to the site, its contours and boundaries, the risk of injury or damage to adjacent property and its occupiers, the supply of and conditions affecting availability of labour and materials, and 'any other matters or information affecting or likely to affect the execution of, or price tendered for, the Works'. It puts the contractor in the same position as if he *had* satisfied himself by inspection of the site etc. Contractors must, therefore, take due precautions and so satisfy themselves by inspection etc.

The exception to this provision is in *clause 7(2)* which covers 'any information given or referred to in the Bills of Quantities which is required to be given in accordance with the method of measurement expressed in' them, which may include, for example, the provision of site information by the Authority. Clause 3(1) says that, unless otherwise stated therein, the bills are deemed to have been prepared in accordance with the method of measurement expressed therein. In some cases this may give rise to a claim by the contractor: see *C. Bryant & Son Ltd v. Birmingham Hospital Saturday Fund* (1938).

Although there are no procedural requirements in the sub-clause, should the contractor be of the view that the bills have not been prepared in accordance with the specified method of measurement, he should notify the PM.

Clauses 7(3) to 7(5) deal with 'unforeseeable ground conditions' which are (clause 1(1) and clause 7(3)): 'ground conditions (excluding those caused by weather but including artificial obstructions) which he [the contractor] did not know of, and which he could not reasonably have foreseen having regard to any information which he had or ought reasonably to have ascertained' and certified as such by the PM under sub-clause (4). It alters the common law position (see *Appleby v. Myers* (1867)).

Unfortunately, the degree of probability that the contractor is required to foresee is not defined. An experienced contractor might foresee the *possibility* of many conditions or obstructions to the work, but it is submitted that he is not required to allow for those where the possibility is too remote, i.e., he would not be expected to foresee and allow for conditions or obstructions that are no more than mere possibilities. There is a considerable body of case law under the corresponding provision (clause 12) in the ICE Conditions of Contract, 5th edition, and reference may usefully be made to *C.J.Pearce & Co. v. Hereford Corporation* (1968) and *Holland Dredging (UK) Ltd v. The Dredging & Construction Co Ltd* (1987) for guidance.

If the contractor becomes aware of unforeseeable ground conditions during the execution of the works then, under *clause 7(3)*, he must inform the PM *immediately* of those conditions, and also state the measures which he proposes to take to deal with them. The contractor's notice must be in writing: see clause 1(3).

It is suggested that the notice must be specific in its terms. The purpose of the notice is not merely to warn the PM that a problem has arisen, but also to give him the opportunity to take remedial action. He might, for example, vary the work so as to avoid an obstruction or suspend the execution of the work (clause 40). (Clause 7(5) assumes that the PM may give relevant instructions and clause 40 is sufficiently wide in its terms to give the PM many options).

On receipt of the contractor's notice, if the PM agrees that the specified ground conditions could not reasonably have been foreseen by the contractor having regard to any information he should have had, he is to certify accordingly. He must notify the contractor (in writing) of his decision *(clause 7(4))*. and any dispute could be referred to adjudication if he and the contractor cannot agree.

Clause 7(5) deals with the financial consequences where unforeseeable ground conditions are encountered. It says that 'without prejudice to any instruction given by the PM' (which will be valued accordingly under clause 42 or 43) if as a result of those conditions the contractor 'properly carries out or omits work which he would not otherwise have carried out or omitted' it is to be valued as a variation under clause 42. The contract sum is then to be increased or decreased as appropriate.

Clause 7(6) provides that no claim by the contractor for additional payment will be allowed because the contractor has misunderstood or misinterpreted anything mentioned in clause 7(1). It further states that the contractor shall not be released from any risks or obligations arising under the contract or because 'he did not or could not foresee *any* matter which might affect or have affected the execution of the Works'.

Clause 8 Insurance

Clause 8 is entirely new. It imposes an obligation on the contractor to maintain relevant insurance for the duration of the contract and for the longest specified maintenance period. There are alternative provisions as to the type of insurance to be maintained; which alternative applies will be specified in the abstract of particulars.

Clause 8(3) Alternative A, deals with the standard 'all risks' insurance which most contractors now maintain. Clause 8(3) Alternative B covers 'Combined Site Insurance' which is taken out in the joint names of the Authority, the contractor and all sub-contractors: see *Petrofina (UK) Ltd v. Magnaload Ltd* (1983). Where this alternative applies, it will be so stated in the abstract of particulars and will be supported by a 'Summary of Essential Insurance Requirements' issued by the PSA. This will be, in effect, a specification for a suitable policy of insurance.

Clause 8(1) requires the contractor to maintain the following insurance policies:

- Employer's liability insurance covering his employees.
- Insurance against loss or damage to the works and things for which he is responsible under the contract: see clauses 13 and 19. Cover must be for 'the full reinstatement value' (including the cost of transit and off-site risks): see clause 8(2).
- Insurance against personal injury to any persons and loss or damage to property (i.e., third party liability) not otherwise covered.

The cover may be by means of either existing or new policies but it must be 'for the duration of the Contract and the longest maintenance period': see clause 39(1) as to the duty of the PM to issue a certificate that the works are in a satisfactory state when the longest maintenance period ends.

The expression 'full reinstatement value' in *clause 8(2)* is important. This will obviously increase in value as the work proceeds and in the contractor's interests the cover must be adequate. The sum insured must reflect the actual cost of both reinstatement and/or any lost or damaged materials. It should also include the cost of removing debris. The term also appears to cover consequential loss in this context, e.g., any loss suffered by the Authority because of delay in completing the works or any increased cost of carrying out work not done at the time the damage occurred: see the definition of 'loss or damage' in clause 19(6).

Reference should also be made to clause 1 for the definition those

General obligations

'accepted risks' which are the Authority's responsibility: see clause 19(5) for the Authority's liability.

Where Alternative A of *clause 8(3)* is applicable, the Authority has the right to receive, on request, a copy of the insurance policies effected or held by the contractor. Within 21 days of acceptance of the tender the contractor must send to the Authority a certificate from his insurer or broker attesting that appropriate insurance policies are in force. He is under a like duty to send a similar certificate to the Authority within 21 days of any subsequent renewal or expiry date of the insurances. The insurer's certificate is to be in the form attached to the abstract of particulars.

Where Alternative B is specified then, in addition to employers' liability insurance (which is compulsory in any event under statute), the contractor must take out and maintain a *joint names policy* in accordance with the 'Summary of Essential Insurance Requirements' attached to the abstract of particulars. This policy must be in the joint names of the Authority, the contractor and all his sub-contractors. The Authority has the right to receive a copy of the policies effected and the contractor must send an insurer's or broker's certificate to the Authority within 21 days of acceptance of the tender.

Clause 8(4) confers default powers on the Authority. If the contractor fails to meet his insurance obligations, the Authority can take out appropriate cover and deduct the cost of doing so from any advances due to the contractor: see clause 48. Other relevant procedural provisions are made.

Clause 8(5) is inserted *ex abundanti cautela* ('from an excess of caution'). Nothing in clause 8 relieves the contractor from any of his obligations and liabilities under the contract.

Clause 9 Setting out

Under *clause 9(1)* it is the duty of the PM to provide such dimensioned drawings, levels and other information as he considers reasonably necessary to enable the contractor to set out the works. This is stated to be 'subject to any express provision in the Contract to the contrary', e.g., a provision in the bills. It is obviously important that the PM should provide this information at such times and in sufficient detail to enable the contractor to proceed expeditiously in accordance with the contract.

The contractor is made responsible for setting out the works. He is also to provide all necessary instruments, profiles, templates and rods. The contractor is made *solely responsible* for the correctness of the setting out. If he finds discrepancies in any setting-out information provided, he should inform the PM as soon as possible.

If the works are set out so that a trespass occurs to neighbouring property, the contractor is responsible and must indemnify the Authority against any liability under clause 19. However, it is thought that if trespass arose because of the Authority's neglect or default, the contractor would have a right to reimbursement under clause 19(5). In any event, in such a case, the contractor could obtain a full indemnity from the Authority under the general law: Civil Liability (Contribution) Act 1978 and *Kirby* v. *Chessum & Sons Ltd* (1914).

Delay in supplying the necessary information may give rise to a money claim under clause 46 (2)(a) and to an extension of time under clause 36(2)(b). It would also amount to a breach of contract at common law: *Neodox Ltd* v. *Borough of Swinton and Pendlebury* (1958).

Clause 9(2) obliges the contractor to provide, fix and be responsible for the maintenance of all stakes, templates, profiles etc. He must take all necessary precautions to prevent them being interfered with or removed and he is made responsible 'for their efficient reinstatement' if they are disturbed.

General obligations 63

Clause 10 Design

Edition 2 of GC/Works/1 did not deal with the problems of design liability. Clause 10 sets out the contractor's design liability in very clear terms and thus avoids the 'creeping' design liability which often occurred under Edition 2, especially in relation to specialist services. Clause 10 recognises the realities of the situation because in the majority of projects it is impossible for the design to have been fully specified at tender stage and on any construction project the contractor in fact must carry out some on-site design work, i.e., 'second order' design. The essential source of reference to this problem area is *Design Liability in the Construction Industry*, 3rd edition, by D.L.Cornes (1989, BSP Professional Books, London).

Clause 10(1) provides that if the contractor 'either by himself or by means of any servant, agent, *sub-contractor or supplier* is required under the Contract to undertake the design of any part of the works' he must submit appropriate design information to the PM for approval. The contractor must not commence any work to which such information relates unless and until the design has been approved in writing by the PM. There is a further restriction: the contractor must not alter the approved design without the further written approval of the PM. The PM's approval does not relieve the contractor of any liability: see clause 10(4).

It is important to note that the contractor's liability includes design work by his sub-contractors, whether nominated or domestic, and the contractor is not relieved of any responsibility where the Authority obtains a direct warranty from a sub-contractor (see clause 10(3)).

Clause 10(2) specifies the standard to be expected of the contractor. His liability is the same as applies to design undertaken by a professional designer. This is essentially an obligation to act without negligence, the position being thus expressed in *Bolam* v. *Friern Hospital Management Committee* (1957) by Mr Justice McNair:

'How do you test whether this act or failure is negligence? In an ordinary case it is generally said that you judge that by the action of the man in the street. He is the ordinary man. In one case it has been said that you judge it by the conduct of the man on the Clapham omnibus. He is the ordinary man. But where you get a situation which involves the use of some special skill or competence, then the test whether there has been negligence or not is not the test of the man on top of the Clapham omnibus because he has not got this special skill. A man need not possess the highest skill at the risk of being found negligent. It is well established law that it is sufficient if he exercised the ordinary skill of an ordinary competent man exercising that particular art'.

As David Cornes so aptly puts it (*op.cit.*, p.51):

'To put it another way, in deciding whether there is negligence in a particular case it is necessary to look at what an ordinary competent designer exercising the particular skill would do and to compare that with the actions of the person against whom the negligence is alleged.'

Clause 10(2) is so important that it must be set out in full. It says:

'The Contractor's liability to the Authority in respect of any defect or insufficiency in any design undertaken by the Contractor himself or by means of any servant, agent, sub-contractor or supplier shall be the same as would have applied to an architect or other appropriate professional designer who had held himself out as competent to take on work for such design and who had acted independently under a separate contract with the Authority and supplied such a design for, or in connection with, works to be carried out and completed by a contractor not being the supplier of the design.'

Despite this broad statement, it is thought that if a performance specification is included, the contractor might well be under the higher obligation of *fitness for purpose.*

Clause 31(2) requires him to use 'the skill and care of an experienced and competent contractor' to see that both the works and 'any things for incorporation, whether or not [he] is required to choose or select any of them *are of good quality for their intended purpose and conform to the requirements of the specification . . . .* '

The contractor is under a duty to notify the PM of any problems which come to his notice, and while this is not a full design check (as has been suggested) it is clear that the contractor must accept some fitness liability.

Clause 10(3) reflects the fact that direct warranty agreements between an employer and specialist sub-contractors and other designers are quite common. It states that the contractor's liability is not affected by any warranty that the Authority may obtain from a sub-contractor.

Clause 10(4) provides that the PM's approval of the contractor's design in no way relieves the contractor of his liability for the design.

Clause 11 Statutory notices

This clause is an amended version of clause 14 in Edition 2. It deals with the giving of statutory notices and closely related matters. It does not expressly bind the contractor to comply with the substantive provisions of the relevant legislation, but as will be seen, he is bound to do so. It is clearly an implied term of GC/Works/1 that the contractor will comply with statute in any event. This was so held in *Street* v. *Sibbabridge Ltd* (1980).

In that case Judge Edgar Fay QC, Official Referee, held that it was an implied term in a building contract that the contractor would comply with the current building regulations and that this overrode the contractor's express contractual obligation to comply with the architect's design specification and instructions about the depth of foundations.

Reference may also be made to *Townsends (Builders) Ltd* v. *Cinema News and Property Management Ltd* (1958) where the Court of Appeal had no difficulty in holding that the contractor was under an obligation to comply with byelaws as well as to give notices required by them by the terms of an ineptly drafted clause in the then RIBA Conditions of Contract.

The substantive provisions (but not the procedural provisions) of the Building Regulations 1985 apply to Crown buildings: Building Act 1984, section 44 (when activated). This will affect defence establishments and various Government offices. Similar provision is made (1984 Act, section 45) for the buildings of the United Kingdom Atomic Energy Authority.

Clause 11(1) imposes on the contractor an obligation to give all notices required by statute, regulations or byelaws and pay any necessary fees or charges, and to supply all drawings and plans required in connection with any statutory or allied notice.

Clause 11(2) requires the contractor to obtain any necessary consents, permissions or licences from any statutory undertakers or adjoining owners whose services or land may be affected by the works or whose consent is necessary in connection with them. He must also pay any consequential fees or charges.

Clause 12 Patents

The principle of this clause is that if royalties, licence fees or other sums are payable for any patented articles, processes, inventions, etc. used for or in connection with the works, these are the contractor's responsibility: *Clause 12(1)*.

Under *clause 12(2)* the contractor gives the Authority an indemnity against any liability arising in connection with the use of such articles.

Clause 12(3) ensures that where liability is imposed on the contractor in respect of patented articles, processes, etc. which are *necessarily* used as a result of compliance with a variation instruction issued under clause 40(2)(a), the contractor will be reimbursed in the valuation of the VI. This is subject to the condition (paragraph (b)) that the use 'was not reasonably contemplated under the Contract'.

Clause 12 is a redrafted version of clause 15 in Edition 2.

Clause 13 Protection of works

This important provision is an extended version of condition 17 of Edition 2 which was merely in the nature of a specification requirement.

Clause 13(1) provides that, during the execution of the works, until completion, the contractor must take all measures and precautions necessary to take care of the site and the works. He 'shall [also] have custody of all Things on the Site against loss or damage from fire and any other cause.' The wording is not particularly happy, but the intent is plain. The Works etc. are at the contractor's sole risk (subject to the limited exceptions in condition 19(5)). The contractor is under corresponding indemnity obligations to the Authority by the terms of clause 19.

Clause 19(5) sets out the exceptions to this general rule and provides that the Authority is to reimburse the contractor where any loss or damage is caused by:

(a) its neglect or default or that of any of its other contractors or agents;
(b) any accepted risk (see the definition in clause 1(1)) or unforeseeable ground conditions (condition 7(3)) or 'any other circumstances which are outside the control of the contractor or any of his sub-contractors or suppliers and which could not have been reasonably contemplated under the contract'.

The second sentence of sub-clause (1) makes the contractor 'solely responsible' for the taking of all reasonable and proper steps for protecting, securing, lighting and watching all places on or about the works and the site 'which may be dangerous to his workpeople or to any other person'.

Contractors will be aware of the statutory obligations imposed on them by the Health and Safety at Work etc. Act 1974 and various regulations in respect of their employees and others, and of their liability under the Occupiers' Liability Acts 1957 and 1984, as amended.

Clause 13(2) states what would in any case be the law, namely that the contractor must comply with any statutory regulations governing the storage and use of anything brought on to the site in connection with the works, and this is so whether or not the relevant regulations bind the Crown.

Clause 14 Nuisance and pollution

Clause 14 is a verbatim repetition of clause 18 in Edition 2. Under it, the contractor must take all reasonable precautions to prevent nuisance or inconvenience to the owners, tenants or occupiers of other properties and to the public generally, which is in any event his obligation under the general law. The contractor is also required to take efficient measures to prevent pollution of streams and waterways. In the current climate of the frequent service of statutory notices and private injunctions to prevent noise, dust and vibration, the observance of this provision is important. It puts beyond doubt the fact that the cost of compliance is on the contractor.

Clause 15 Returns

This provision follows closely the wording of clause 34 of Edition 2 and its effect is the same. The contractor's agent (see clause 5) must provide the PM with daily returns showing the number and description of workpeople and the plant employed on the works. The return is to be in the form required by the PM.

Clause 16 Foundations

Closely following the wording of clause 21 in Edition 2, this short clause provides that the contractor must not lay foundations until the excavations for them have been examined and approved by the PM. If he does so, he is in breach of contract. The PM, of course, has the necessary power (clause 40(2)(d)) to reject and require removal of defective work. In light of the alarming number of cases involving failure of foundations which come before the courts, this provision is of immense practical importance.

Clause 17 Covering work

This clause is virtually a verbatim repetition of the provision found in clause 22 of Edition 2. It imposes on the contractor an obligation to give the PM reasonable notice whenever any work or materials are intended to be covered up. The notice must be in writing: see clause 1(3). What length of notice is reasonable will depend on all the circumstances of the case. The provision could be of importance in the case of an action which was alleged to be statute-barred and the Authority the pleaded 'deliberate concealment' under section 32 of the Limitation Act 1980: see pp.28–29, *ante.*

If the contractor fails to give due notice prior to covering work etc., the PM can require him to uncover the work at his own expense. Despite the use of the phrase 'with work or otherwise' this provision would not appear to be limited to excavations as has sometimes been alleged.

Clause 18 Measurement

Clause 18 is modelled on clause 37(1) in Edition 2, though its wording has been modernised and the contractor's obligations are now sub-divided logically. In fact, because of the new provisions in clause 42 which empower the PM to require a lump sum quotation from the contractor in respect of variations, the need for the measurement and valuation of variations for the preparation of the final account should be more limited. However, there is still need for on-site measurement.

Clause 18(1) requires the attendance of the contractor's representative 'from time to time', on reasonable notice from the quantity surveyor, to attend at the works and to take jointly with him any measurements of the work executed that may be 'necessary for the preparation of the Final Account'. The quantity surveyor's notice must be in writing: clause 1(3). The measurements and any differences in relation to them are to be recorded in the manner required by the quantity surveyor. The sub-clause recognises the usual practice of the contractor's quantity surveyor attending jointly with the employer's quantity surveyor to take measurements.

Clause 18(2) requires the contractor to provide the appliances and other things necessary for measuring the work. He must do this at his own expense.

Clause 18(3) says that if the contractor's representative fails to attend when required under sub-clause (1), the quantity surveyor may take the measurements by himself. In that event 'those measurements shall for the purpose of the Final Account be final and conclusive'.

Clause 19 Loss or damage

This clause deals with the contractor's liability for any loss or damage arising out of or in any way connected with the carrying out of the contract works. It reinforces the generally absolute nature of the contractor's obligation to complete and his liability for the works for the duration of the contract. However, the contractor's liability is far more extensive than that. It extends to all kinds of loss and damage to people and property, including third party claims. (See clause 8(1) for the corresponding obligation to insure.)

Clause 19(1) sets out the scope of the provision. It applies to 'any loss or damage which arises out of or [which] is in any way connected with the *execution or purported execution* of the Contract'. 'Loss or damage' is defined (clause 19(6)) for the purposes of the clause as including:

- Loss or damage to property
- Personal injury to or sickness or death of any person
- Loss or damage to the works or any thing not for incorporation which is on the site; and
- Loss of profits or loss of use suffered because of any loss or damage

This definition could scarcely be wider.

Under *clause 19(2)* the contractor's obligation is specified. He must, without delay and at his own cost, 'reinstate, replace or make good to the satisfaction of the Authority ... any loss or damage'. Alternatively, if the Authority so agrees, he must 'compensate the Authority' for any loss or damage.

Subject to what is said in clause 19(5) below, the contractor's obligation is an absolute and unqualified one, and the extensive nature of a similar obligation was considered in *A.E.Farr Ltd* v. *The Admiralty* (1953), which arose under an earlier edition of what is now GC/Works/1, Edition 3. In that case it was held that the Authority was entitled negligently to knock down what the contractor had erected and then require him to rebuild at his own expense. (The works in question were a jetty which was damaged by a destroyer, owned by the Authority, which collided with it.) The clause in that case referred to damage arising 'from any cause whatsoever'.

Of course, under Edition 3, the result would be different because the contractor would be able to recover in such circumstances under clause 19(5). However the principle involved is the same since liability under the clause is in no way dependent on the contractor's fault. The only exceptions to this basic rule are set out in clause 19(5), which is discussed below.

Clause 19(3) deals with third party claims brought against the Authority. Under it, where a claim is made and proceedings are brought against the Authority for any loss or damage as defined, the contractor must reimburse the Authority for 'any costs or expenses which the Authority may reasonably incur in dealing with, or in settling,' the claim or proceedings.

Clause 19(5) sets out three cases in which the Authority must reimburse the contractor for any costs or expenses he has incurred under clause 19(2) and (3). These are:

- Where the loss or damage is caused by the neglect or default of the Authority or any of its other contractors or agents: paragraph (a).
- Where the loss or damage is caused by an accepted risk or by unforeseeable ground conditions.

'Accepted Risks' are defined in clause 1(1) as meaning the risks of:

(a) pressure waves caused by the speed of aircraft or other aerial devices
(b) ionising radiations or contamination by radioactivity from any nuclear fuel or from nuclear waste from the combustion of nuclear fuel
(c) the radioactive, toxic, expolosive or other hazardous properties of any explosive nuclear assembly (including any nuclear component); and,
(d) war, invasion, act of foreign enemy, hostilities (whether or not war has been declared), civil war, rebellion, insurrection, or military or usurped power.

'Unforeseeable ground conditions' are defined in clause 1(1) by reference to clause 7(3).

- Where the loss or damage is caused by any other circumstances outside the control of the contractor, his sub-contractors or suppliers which could not have been reasonably contemplated under the contract.

It should be noted, however, that under clause 13(1) the contractor is bound to 'take measures and precautions needed to take care of the Site and the Works . . . . against losses or damage by fire and any other cause', and the Authority's obligation to reimburse the contractor is '*to the extent that the loss or damage is caused by*' the things specified in paragraphs (a) to (c).

Clause 20 Personal data

This new provision relates to the disclosure or misuse of information about private individuals which is held in a computer or other electronic system. 'Personal date' is given the same meaning as in section 1(3) of the Data Protection Act 1984. The essence of the provision is that it imposes an obligation on the contractor to reimburse the Authority for all losses, costs and expenses which it incurs as a result of claims and proceedings for disclosure or misuse of personal data. This is an absolute liability and liability is not conditional on the contractor having been negligent.

Clause 21 Defects

Clause 21 deals with defects which appear *after* completion and during the maintenance period specified in the abstract of particulars. Defects which arise before completion are, of course, dealt with under clause 31.

Post-completion defects are dealt with in a particularly effective manner since clause 21(1) obliges the contractor to make good without delay and at his own cost 'any defects in the Works resulting from what the Authority considers to be default by the contractor' or those for whom he is responsible in law.

It is only after the contractor has completed the remedial works that any question of reimbursement arises. If the contractor then establishes that the defect is not his responsibility then the employer undertakes to pay him the cost of the remedial works.

Clause 21(1) provides that the contractor must make good any defects in the works notified to him by the Authority and which appear during the relevant maintenance period. The defects must result from what the Authority believes to be default by the contractor or any agent or subcontractor of his. The contractor must make good such defects at his own expense and without delay. No definition of 'defect' is given. Furthermore, the clause is silent as to how long after the specified maintenance period the contractor may be notified; the only requirement is that the defects must *appear* within the relevant maintenance period. It is suggested, therefore, that he remains liable to be notified for the whole of the limitation period: see p.28.

Clause 21(2) sets out the circumstances in which the contractor is entitled to payment for remedial works. The Authority undertakes to reimburse the contractor after completion of the remedial works for those costs which he has incurred to the extent that he satisfies the Authority that any defects were not caused by:

(a) his neglect or default or that of one of his agents or subcontractors; or
(b) by any circumstances within his or their control.

This is not a substantive change but only a shift in the burden of proof; it is for the contractor to establish, on the balance of probabilities, that the defect is not his responsibility.

Under *clause 21(3)* if the contractor fails to make good any notified defect, the Authority is entitled to carry out the work and charge the costs and expenses to the contractor. The Authority may do this itself, using its own employees, or use other contractors. If the contractor fails to comply with his obligation to remedy defects, the Authority may employ others to do so and recover any costs and expenses involved from the contractor.

Clause 21(4) is important. It provides that, in the case of any defects which have been made good and which are subject to a separate sub-contract maintenance period, the defects liability provisions apply in that case until whichever is the later of:

(a) the expiry of either the appropriate sub-contract maintenance period, or
(b) six months from the date of making good.

Clause 22 Government premises

This clause applies where work is being executed 'within the boundaries of Government premises', e.g., Admiralty or other service establishment. It follows closely the wording of Edition 2, clause 35, and requires the contractor to comply with the rules and regulations of the occupying department or body, and with such amendments to them as may be notified during the execution of the works.

Clause 23 Racial discrimination

The first part of this clause (sub-clause (1)) makes contractual what is already a statutory obligation, namely that the contractor shall not unlawfully discriminate on racial grounds contrary to the provisions of the Race Relations Act 1976 or any statutory modification or enactment of it.

By the terms of sub-clause (2) the contractor must take all reasonable steps to ensure that there is no such discrimination by his employees, agents or sub-contractors.

Clause 24 Corruption

So important is this provision that breach of it is listed under clause 56(6)(f) as a ground on which the Authority may determine the contract and it mirrors the provisions of clause 55 in Edition 2, which is echoed in sub-clause (2) of this clause.

The provision is self-explanatory. It prohibits the contractor in the widest possible terms from indulging in corrupt practices. Any breach of the condition entitles the Authority to determine the contract and to recover from the contractor the amount of the bribe etc.: see *Reading* v. *Attorney-General* (1951).

Such practices are, in any event, a criminal offence under the Prevention of Corruption Acts 1889 to 1916 and the clause merely makes the legal position explicit to the contractor. It is a matter of regret that the decline in moral standards has led to a great number of these offences within the construction industry in recent years.

The final sub-clause (clause 24(1)) makes the decision of the Authority as to whether there is a breach of this provision 'final and conclusive'.

Clause 25 Records

This is a new provision which reflects the importance of records to efficient management of the contract, especially in regard to the making and settlement of any claims.

Clause 25(1) obliges the contractor to keep, for the purposes of the contract, 'such records as may be necessary for the QS, the PM or the Authority to ascertain or verify any claims made . . . . or any sums to be paid to the contractor . . .' It merely makes a matter of contractual obligation what should already be good contract practice. A not dissimilar obligation is, of course, imposed by clause 52(4) of the ICE Conditions of Contract for Civil Engineering Works, 5th edition.

The form of the records is not specified, but they must be 'such . . . as may be necessary for the QS, the PM or the Authority to ascertain or any claims' and, practically, it is desirable that the sort of records required to be kept should be agreed with the PM and the QS, preferably before the contract commences.

Clause 25(2) states that the contractor must allow the PM and quantity surveyor access to these records so that they may discharge their contractual functions. He must also supply them with the information they require 'including means to interpret the records'.

Chapter 5

Security

Introduction

The four clauses in this section of the form all derive from earlier editions and result from the security aspects attending most Government contract work. Should GC/Works/1, Edition 3, be used for private sector work, then they would need to be deleted from the contract as they have no relevance at all to private employers. The importance of security on Government projects no doubt accounts for the fact that they are given prominence in the body of the contract. It is not thought that any detailed commentary or explanation is called for.

Clause 26 Site admittance

Special security considerations apply to many Government projects and sites and the security of the State is and must be paramount. This is the reason for this provision which enables the Authority, through the PM, to exercise control over the people who may be admitted to site. It is based on clause 56 in Edition 2.

Clause 26 should be read in conjunction with clause 6 (contractor's employees) and with clause 56 which gives the Authority a power of determination for breach of this provision (clause 56(6)(e)) where the Authority determines that the contractor's failure is prejudicial to the interests of the Crown. Under Clause 56(9) such a decision of the Authority is stated to be 'final and conclusive'.

Clause 26(1) requires the contractor to take reasonable steps to prevent the admission of unauthorised persons to the site (as defined in clause 1(1)). The PM may give the contractor notice that any person is not to be admitted to the site. The notice must be in writing: see clause 1(3). Where notice is so given, the contractor must take all reasonable steps to prevent the person being admitted to the site.

Under the terms of *clause 26(2)* the contractor must provide the PM with a list of the names and addresses of those who are or may at any time be concerned with the whole or part of the works, and the capacities in which they are so concerned. He must also provide such other particulars as the PM reasonably requires. The purpose of this provision is, of course, different from that of Clause 6(1), under which the PM may require the contractor to keep undesirable employees off the site.

Clause 26(3) makes the decision of the PM final and conclusive on matters arising under clause 26 and this, and the provisions of clause 56 enabling the Authority to determine the contract for breach of this provision (and in particular clause 56(9) which makes the decisions of the Authority under clause 56 'final and conclusive') are justified by the need to exclude matters involving security considerations from scrutiny in arbitration or litigation.

Clause 27 Passes

No particular comment is required on this clause which is a redrafted version of clause 57 in Edition 2. It provides for the issue by the PM of admission passes for the contractor's employees in appropriate cases. Passes are normally required for entry to Government establishments. There is no express sanction if the contractor fails to return the passes as required by the last sentence and it is submitted that such failure would amount to no more than a technical breach of contract.

Clause 28 Photographs

The effect of this short clause is that the contractor must not take any photographs of the site or of the whole or part of the works without the prior written consent of the PM. The contractor is also obliged to take 'all reasonable steps' to see that no such photograph is taken, published or otherwise circulated by any of his employees unless such prior consent is obtained. It is almost word for word the same as clause 58 in Edition 2.

Clause 29 Secrecy

It is not thought necessary to discuss this clause in detail. It derives from clause 59(1) of Edition 2.

Briefly, it draws the contractor's attention to the provisions of the Official Secrets Acts 1911 to 1939 and section 11 of the Atomic Energy Act 1945, and imposes on him an obligation to take reasonable steps to ensure that these statutory provisions are brought to the attention of his employees and sub-contractors. It also emphasises that information obtained by the contractor about the contract etc. is confidential and non-disclosable and is to be treated as such. This sort of obligation of confidentiality often arises in contract, and the most common remedy for breach of confidence is the injunction. Damages may be awarded where loss resulting from an actual breach is suffered.

Chapter 6
Materials and workmanship

Introduction

One of the core obligations of the contractor — namely the manner in which he is to execute the work — is contained in this fourth section of the form, curiously sandwiched between a vesting clause and a provision dealing with materials arising from excavations and fossils, antiquities and other objects of interest or value.

At common law, in all lump sum types of building contract, the contractor's obligation is to complete the defined work on or before the stipulated date for completion. Unless there is contrary provision, the contractor implicitly agrees that he will do his work in a good and workmanlike manner, using good and proper materials, and that the completed structure will be reasonably fit for its intended purpose: see *Hancock* v. *B.W.Brazier (Anerley) Ltd* (1966). The third limb of the duty will not normally arise where the building owner employs an architect: *Lynch* v. *Thorne (1956)*

Similar duties arise in the case of design and build contracts: see *Greaves & Co. (Contractors) Ltd* v. *Baynham Meikle & Partners* (1975) and *Surrey Heath Borough Council* v. *Lovell Construction Ltd* (1988). In the later case it was held that there is an implied term in a JCT 81 design and build contract that the contractor would carry out both his design and construction functions in a good and workmanlike manner and with the proper care of a skilled contractor.

Well-drafted construction contracts will, as GC/Works/1, Edition 3 does, particularise the contractor's obligations in greater detail, not only as regards his duty to complete (see clause 34(1) but also as to the manner in which he is to execute the works (see clause 31(2)).

The general principle applicable to building and engineering contracts is that, in the absence of any indication to the contrary, a contractor is entitled to plan and perform his work as he pleases, provided always that he finishes it by the time fixed in the contract. This is shown by *Wells* v. *Army & Navy Co-operative Society* (1902) where Mr Justice Wright said:

'The plaintiffs were entitled to do the work in what order they pleased'.

In *Greater London Council* v. *Cleveland Bridge & Engineering Co. Ltd* (1986), which arose out of the construction of the Thames Barrier, the employers contended for the implication of a term that the contractors should exercise 'due diligence and expedition' in the performance of the contract and alleged that the contractors had failed to comply with that term even though they had completed the works by the due date. Both Mr Justice Staughton and the Court of Appeal rejected the employer's contention and the contrary view expressed in *Hudson's Building and Engineering Contracts*, 10th edition, p.611 (1970, Sweet & Maxwell, London).

Clauses 31(1) and 34 address this problem directly by imposing an express subsidiary obligation on the contractor requiring him to 'proceed with diligence and in accordance with the Programme ... so that the whole of the Works ... shall be completed ... by the Date ... for Completion': clause 34(1). Whether breach of that obligation would sound in substantial damages is a doubtful question, and the Authority's more effective remedy is determination of the contract under clause 56. This confers on the Authority a right to dispose of a dilatory contractor who is failing to meet his subsidiary obligations.

Clause 30 Vesting

Clause 30 is a traditional vesting clause transferring the property in the contractor's plant and materials to the Authority during the currency of the contract. It is a redrafted version of clause 3 in Edition 2. Its objective is to improve the position of the Authority in the event of a failure by the contractor to complete the contract, in particular where the failure is caused by the contractor's insolvency. It is plainly effective to defeat claims made by the contractor's liquidator or any judgment creditor until the contract is completed. The position as regards vesting clauses is not free from doubt e.g., it is arguable that in a true individual bankruptcy situation the trustee in bankruptcy might be able to defeat the Authority's interest (*Re Fox, Oundle and Thrapston RDC* (1948)). However, it is the writer's view that the present clause is effective subject to the limitations discussed below.

There is an important difference between those items which are brought on site intended permanently to become the property of the Authority, i.e., materials intended for incorporation in the works and called 'Things for incorporation' in clause 1(1), and those things which will eventually be removed from site, such as plant. Both categories are vested in the Authority by the clause ('The Works and any Things on the Site in connection with the Contract . . .'). In the latter case the transfer will only be temporary, and the reference to 'the Works' is really otiose in the majority of cases, because once materials become affixed to the land, i.e., they are built into the works, they become part of the freehold. If the Authority enjoys a lesser estate or interest e.g., leasehold, it enjoys the property during its term (*Tripp* v. *Armitage* (1839); *Elwes* v. *Maw* (1802)) and the same principle is applicable: *Quicquid plantatur solo, solo cedit* ('whatever is affixed to the soil belongs to the soil').

The principal objects of a vesting clause are:

'to provide security to the employer for money advanced and to enable the employer to obtain speedy completion of the works by another contractor in the event of the original contractor's default, by providing materials and plant on site ready to use free from the claims of the original contractor, and his creditors or his trustee in bankruptcy or liquidator': D. Keating, *Building Contracts*, 4th edition (1978, Sweet & Maxwell, London) p.125.

The clause under consideration is particularly tightly drafted. *Clause 30(1)* provides that:

Materials and workmanship

'The Works and any Things on Site in connection with the Contract which are owned by the Contractor or by any company in which the Contractor has a controlling interest, or which vest or which will vest in him under any contract, shall become the property of and vest in the Authority.'

This provision is clearly effective to transfer ownership to the Authority of plant, materials, etc., which are the property of the contractor and would doubtless defeat claims by any third party claiming through the contractor such as a liquidator or trustee in bankruptcy, as numerous cases such as *Re Waugh, ex parte Dickin* (1876) establish. A corresponding clause is to be contained in any sub-contract: clause 62(2). The provision would not, however, be effective as against third parties where the contractor has in fact no title to pass to the Authority and it does not attempt to transfer title in these circumstances.

The common situation is that a supplier of materials sells on the basis of a retention of title clause, i.e., an express stipulation that ownership of the goods shall not pass to the intending purchaser until he has paid for them: *Aluminium Industrie Vaassen BV* v. *Romalpa Aluminium Ltd* (1976). In that case the maxim *nemo dat quod non habet* applies and the contractor could not give title to the Authority if the supplier had retained title until payment. The effectiveness of such provisions has been upheld time and again by the courts. A comprehensive discussion of the legal complexities is to be found in *Effective Retention of Title Clauses* by John Parris (1986, BSP Professional Books, London). Ownership passes to the Authority of course, as soon as the goods are built in to the works even if there is a retention of title clause in the contractor's supply contract: *Reynolds* v. *Ashby* (1904).

The position would be the same as regards the materials on site belonging to a sub-contractor, although clause 62 attempts to deal with this problem by providing that in every sub-contract entered into by the contractor a clause shall be included (clause 62(2)(b)):

'to the effect that from the commencement to the completion of the sub-contract all Things belonging to the person who enters into the subcontract shall vest in the Contractor subject to any right of the Contractor to reject the same.'

If the sub-contract includes such a vesting clause it will be effective, but if it does not then any unfixed materials will remain the sub-contractor's property and cannot vest in the Authority under this clause: *Dawber*

Williamson Roofing Ltd v. *Humberside County Council* (1979).

Clause 30(1) purports to vest in the Authority not only things belonging to the contractor but also those owned 'by any company in which the Contractor has a controlling interest'. This is of doubtful legal effect because a limited liability company has a separate legal personality and it is a fundamental principle of English law (though not of the law of Scotland) that only the parties to the contract can acquire rights and liabilities under it. This doctrine of privity of contract would seem to prevent property belonging to a third party, i.e., those companies in which the contractor has a controlling interest, from vesting in the Authority under sub-clause (1) because those companies are separate legal persons and are not parties to the contract.

Indeed, in the *Dawber Williamson* case it was argued unsuccessfully that a clause in the main contract which transferred title in unfixed materials to the employer on payment was ineffective to defeat the unpaid sub-contractor's rights because the sub-contractors were not parties to the main contract. Mr Justice Mais also rightly rejected the specious argument that a provision in the sub-contract form providing that 'the sub-contractor shall be deemed to have knowledge of all the provisions of the main contract ...' in some mysterious way incorporated the main contract provisions into the sub-contract. In the absence of a term to the contrary, materials brought on site by the sub-contractor remain the property of the sub-contractor until fixed: *Tripp* v. *Armitage* (1839) and *Seath* v. *Moore* (1886) where Lord Watson said:

> 'Materials provided by the builder and portions of the fabric, whether wholly or partially finished, although intended to be used in the execution of the contract, cannot be regarded as appropriated to the contract or as "sold" unless they have been affixed to or in a reasonable sense made part of the corpus.'

Clause 30(1), therefore, is effective to transfer title in 'things on site' (see the definition in clause 1(1)) which are owned by the contractor or whose title is vested in him under any contract, but does not avail against third parties with a prior title to them.

The transfer of ownership in 'things on site' is temporary (see clause 30(4)), and although the sub-clause transfers ownership to the Authority, the risk of loss or damage etc. remains with the contractor, subject to clauses 19 (loss or damage) and 65 (other works).

Clause 30(2) makes clear that in principle the risk of loss and damage is that of the contractor, though this is said to be subject to the provisions

Materials and workmanship

of clauses 19 and 65. Under clause 19 (loss or damage) the contractor is made responsible for 'any loss or damage which arises out of or is in any way connected with the execution or purported execution of the Contract'. However, the contractor is enitled to reimbursement for any losses he suffers to the extent that the loss or damage is caused by the neglect or default of the Authority or any of its other contractors or agents or by any accepted risk (defined in clause 1(1)) or by unforeseeable ground conditions (defined in clause 1(1) by reference to clause 7).

Clause 65(2) relieves the contractor of responsibility for damage done to other works being executed on site unless it is caused by the negligence, omission or default of his workpeople or agents. Any damage done to the works in the execution of other works by the Authority's licensees is also made the responsibility of the Authority.

Under *clause 30(3)* the contractor is made responsible for the protection and preservation of the works and any 'things on site' until completion (as certified by the PM under clause 39(1)) or determination of the contract under clause 56(1). Because this risk is his it is important that adequate and appropriate insurance cover be obtained: see clause 8(1) as to the contractor's obligation to effect insurance, *inter alia*, 'against loss or damage to the Works and Things for which the Contractor is responsible . . .'

Clause 30(4) provides that materials, plant etc. brought on to the site shall not be removed before completion without the consent of the PM. At any time during the progress of the works the PM may instruct or permit the contractor to remove any unused materials etc. or any which have been rejected by the PM. The contractor must then remove them 'forthwith', i.e., as soon as reasonably can be (*London Borough of Hillingdon* v. *Cutler* (1968)) at his own expense and the property in the items so removed revests in him. The PM's instruction or consent must be in writing: clause 1(4). It should be noted that, under clause 40(2)(c) the PM is expressly empowered to order 'the removal from the Site of any Things for incorporation and their substitution with any other Things'.

Under Edition 3, in contrast to the position under clause 3(2) of Edition 2, the decision of the PM as to whether or not he will permit unused materials, plant etc. to be removed from site during the progress of the works is not made final and conclusive.

Clause 31 Quality

This important condition spells out the basic obligations of the contractor to build the works with proper skill and care and it covers a variety of closely related matters. It contains a number of novel features. The contractor must execute the works 'using the skill and care of an experienced and competent contractor' and the clause must be read in conjunction with clause 34 as to commencement and completion. The obligations are expressed more widely than their predecessors in Edition 2 (clauses 7 and 13) and in effect revive the 'duty to warn' which was held to exist at common law in *Equitable Debenture Assets Corporation Ltd* v. *William Moss Group Ltd* (1984) and *Victoria University of Manchester* v. *Hugh Wilson and Lewis Womersley and Pochin (Contractors) Ltd* (1984), although later cases such as *Tai Hing Cotton Mill Ltd* v. *Liu Chong Hing Bank Ltd* (1986) cast doubt upon it, if they did not actually deny its existence. See also *University of Glasgow* v. *William Whitfield* (1988).

Clause 31(1): As we have already seen, under the general law, in the absence of an express term, an obligation as to workmanship and materials will be implied. There is a plethora of case law dealing with these implied terms. In particular there are three oft-quoted decisions of the House of Lords, namely *Young & Marten Ltd* v. *McManus Childs Ltd* (1969); *Gloucestershire County Council* v. *Richardson* (1968); and *Independent Broadcasting Authority* v. *EMI Electronics Ltd and BICC Construction Ltd* (1980), all of which repay careful study. There is a full discussion of the whole subject in Emden's *Building Contracts and Practice*, 8th edition, vol. 1, pp.181–189 (1980, Butterworths, London).

In addition to any other express obligations in the contract, such as those in clause 34(1), and to any terms which might otherwise be implied by law, clause 31(1) imposes four separate obligations on the contractor. They are to execute the works:

- 'with diligence'
- 'in accordance with the Programme'
- 'with all proper skill and care', and
- 'in a workmanlike manner'.

If the contractor does not execute the works in accordance with this four-fold obligation he is in breach of contract but, as already indicated, his failure to execute the works 'with diligence [and] in accordance with the Programme' would probably not sound other than in nominal damages provided he completed the works to the satisfaction of the PM in the manner envisaged in clause 34(1). This is because a clause of this sort, which in effect requires him to proceed with diligence and expedition, is a subordinate and subsidiary obligation to the primary obligation. However, breach of the subsidiary obligation would entitle the Authority to determine the contract under clause 56(1): *Greater London Council* v. *Cleveland Bridge & Engineering Co. Ltd* (1986), CA.

Workmanship is skill in carrying out the task, and although the duty is usually expressed as one to carry out the work 'in a good and workmanlike manner' there is surprisingly little postive case law guidance. Most of the reported cases are old and, to modern eyes, give the obvious answer. For example, in *Pearce* v. *Tucker* (1862) the defendant contracted with the plaintiff to instal a new kitchen range with an old boiler behind. The flues were inefficient and the boiler did not provide hot water. When the plaintiff complained, the defendant said that there was insufficient space to make the flues efficient. The plaintiff replied that had he known of this he would not have ordered the work to be done. Not surprisingly, it was held that the defendant was under a duty to advise the plaintiff that the work could not be done and the plaintiff was held entitled to recover from him the costs of fitting up the range 'in an improper and unworkmanlike manner'.

The 'workmanlike' standard means no more, it is suggested, than it must be of a standard that an employer could expect of an ordinarily skilled and experienced contractor of the type the employer has chosen to employ, but regard must be had to any relevant claims which the contractor has made about his level of competence. It is also thought that, as a matter of general law, the contractor warrants that those whom he employs on the works will have the skills to be expected of their respective trades: see, for example, *King* v. *Victor Parsons & Co. Ltd* (1973); *Hancock* v. *Brazier (Anerley) Ltd* (1966); and *Lynch* v. *Thorne* (1956). Under GC/Works/1, Edition 3, of course, the contractor expressly undertakes in the next sub-clause that he will 'use the skill and care of an experienced and competent contractor', but this seems to be no more than an express statement of the common law position.

For example, in *Worlock* v. *S.A.W.S. and Rushmoor Borough Council* (1981), one David Hicks was a carpenter by trade but he had decided to engage in building contracting work and traded as S.A.W.S. He contracted to erect a bungalow at Farnborough, Hants, for Mrs Dolores Worlock on a labour-only basis in accordance with plans prepared by an architect. Mr Hicks quoted for footings 3 ft 3 inches deep which, in the event, proved totally inadequate because of the ground conditions. In subsequent proceedings, Hicks was held liable to Mrs Worlock for breaches of his contractual obligation to exercise generally over the work the standard of care to be expected of a reasonably competent building contractor.

As Mr Justice Woolf put it:

'In my view, the law is clear. He held himself out as a building contractor. He must therefore be judged as such because it was as a building contractor he was engaged by the plaintiff, albeit on a labour-only basis, and in my view, he was required to exercise generally over the work upon which he was engaged that standard of care which would be expected from a reasonably competent building contractor.'

Clause 31(2) is of great significance. The contractor warrants that he will use the skill and care of an experienced and competent contractor to ensure that both the works *and* any 'things for incorporation', i.e., materials, goods, components etc., are of good quality for their intended purpose, and that they will conform to the requirements of the specification, the bills of quantities and the drawings. On one view this is a novel obligation and in effect the contractor is required to check 'things for incorporation' and when actually assembling or incorporating equipment or materials into the works. Certainly, he must take reasonable steps to ensure that 'things for incorporation' are reasonably fit for their intended purpose.

So important is this obligation that clause 31(2) must be quoted verbatim:

'The Contractor shall use the skill and care of an experienced and competent contractor to ensure that
  (a) the Works,
  (b) *any Things for incorporation, whether or not [he] is required to choose or select any of them,* are of good quality for their intended purpose, and
  (c) conform to the requirements of the specification, the Bills of Quantities and the drawings.'

The Authority relies on him when checking things for incorporation and when he is actually assembling or incorporating materials or equipment into the works. His obligation is to take *reasonable care* in exercising his responsibilities and to notify the PM of any problems that come to his notice. This is not a full design check, but it is clear that the contractor must accept some fitness liability. On balance, however, it seems that if there is to be liability a reasonably competent contractor would have to have appreciated that the things were not to standard. If, for example, the contractor becomes aware of a lack of buildability in the design, then he is under a duty to warn the PM under clause 31(3). Readers should study the two seminal judgments of Judge John Newey QC, Official Referee, in *Equitable Debenture Assets Corporation Ltd* v. *William Moss Ltd* (1984) and *Victoria University of Manchester* v. *Hugh Wilson and Lewis Womersley and Pochin (Contractors) Ltd* (1984) for some indication of the extent of the obligation and should disregard the alarmist views already expressed in some quarters.

The sub-clause does not impose on the contractor an obligation to undertake a comprehensive check of the design supplied to him, but he must accept responsibility for detecting potential problems as he handles the design information or as his employees and sub-contractors execute the works. In the *William Moss* case, for example, nominated sub-contractors fixed curtain-walling, which was defective both in design and workmanship, and the building leaked from an early stage. Judge John Newey QC said that the main contractors, Moss, had undertaken the building of the office block:

'... which they performed by their own workmen and sub-contractors. At the very least Moss's staff must I feel certainly have seen what was happening on site, including the activities of Lux-Fix's fitters. In my view, because of the amount of curtain walling that had to be fixed, Moss's staff must have become aware of the difficulty of applying sealants in accordance with [the] drawings; in other words they must have known of the lack of buildability in the design.'

Clause 31(3) says that, in accordance with his duty under clause 31(2)) the contractor must notify the PM before he incorporates any 'things for incorporation' which he considers should not be incorporated because, for example, they are manifestly unfit for their intended purpose. This should be compared with the contractor's obligation in clause 2(3) to inform the PM of discrepancies and internal inconsistencies in the specification and drawings. It is a contractual duty to warn in the *Moss* sense.

Clauses 31(4) to 31(7) follow closely provisions formerly found in clause 13 of Edition 2, although the wording has been clarified.

Under *clause 31(4)* if the PM so requests, the contractor must demonstrate to the PM that he is performing the duties imposed on him by the previous sub-clauses. It also gives the PM power to inspect and examine any part of the works or to inspect, examine and test materials, components and so on, either on the site or at the place where they are being constructed or manufactured or at any place where they are lying or from which they are being obtained. This power may be exercised 'at any time'. The contractor must give the PM 'the assistance and facilities he may reasonably require' in connection with his inspection and examination. This sub-clause envisages that the PM may exercise the right to test and inspect goods etc. on premises which belong to third parties and consequently an appropriate right of access should be reserved in any supply contract or sub-contract. In the case of sub-contracts, this is one of the terms that must be included by virtue of clause 62(2)(c).

The final sentence of clause 31(4) confers on the PM the usual power to reject any things for incorporation which do not conform to the contract documents or which, even though they so conform, are in his opinion of bad quality or which are unfit for their intended purpose.

Clause 31(5) deals with testing by an independent expert to ensure conformity. The decision to arrange for such testing is a matter of the PM's discretion. The costs incurred by the Authority in arranging for tests by an independent expert are borne by the contractor if the test discloses that the goods etc. do not *substantially conform* with the contract requirements. The contractor is also to bear the cost of any *'further tests required to monitor quality following negative test results'*. The report of the independent expert as to fitness or suitability of any materials etc. is made 'final and conclusive' *clause 31(6)*.

Under the terms of *clause 31(7)* if,

- the works or any part of them which do not conform with the contract *and* are not to the satisfaction of the PM; and,
- any things for incorporation which do not conform with the contract and which have been rejected by the PM,

they are to be replaced, rectified or reconstructed at the contractor's expense.

Breach of clause 31 can give rise to determination under clause 56.

Clause 32 Excavations

This provision mirrors clause 20 of Edition 2 and deals with two separate but related matters, namely:

- material and objects arising from excavations; and
- fossils, antiquities or objects likely to have interest or value found on site.

Clause 32(1) provides that all materials and objects of any kind obtained from excavations are to remain the property of the Authority. Under clause 40(2)(l) the PM is empowered to issue instructions regarding the use or disposal of material obtained from excavations on site and under clause 40(2)(m) he has like power in respect of the action to be taken following the discovery of fossils etc. Such an instruction might, dependent on what was ordered, rank as a variation, i.e., if it involved 'any alteration or addition to or omission from the Works or any change in the design, quality or quantity of the Works' (clause 40(5)).

Clause 32(2) can be important if the Authority allows its property, e.g., materials arising from excavations, to be used in substitution 'for any Thing ... which the Contractor would otherwise have provided'. The quantity surveyor is then to ascertain the amount of any saving and the contract sum is to be reduced by its amount: see also clause 48(5) — credit for old materials.

Clause 32(3) provides that 'objects which are or appear to be fossils, antiquities or likely to have value found on Site or in carrying out excavations ...' are the property of the Authority. It is thought that the 'objects .... likely to have value' must be construed *ejusdem generis,* i.e., treated as referring to matters of the same class as 'fossils' and 'antiquities'.

It must be understood that the sub-clause deals only with claims between the contractor and the Authority and it cannot affect any third party's claim to ownership. Items found might well fall within the definition of treasure trove in which case they would belong to the Crown, although the Crown may make an *ex gratia* payment to the finder. Treasure trove is gold or silver coin, plate or bullion or other valuable items hidden in a house or in the earth or other secret place, the true owner being both unknown and undiscoverable. If the property is merely lost or abandoned then it is not treasure trove and in that event it would vest in the Authority under the terms of this sub-clause. 'Fossils', according to *The New Collins English Dictionary*, 2nd edition, (1986

Collins, London) are 'a relic or representation of a plant or animal that existed in a past geological age, occurring in the form of mineralized bones, shells, etc.', while antiquities are ancient relics of any kind.

As soon as any object within the meaning of clause 32(3) is found the contractor is bound forthwith to:

- take all practicable measures not to disturb the object
- cease work if the continuance of the work would endanger or disturb the object or prevent or impede its excavation or removal
- take all necessary steps to preserve the object in its exact position and in the condition in which it was found; and
- inform the PM of discovery of the object and its precise location.

Since the object is made the property of the Authority by clause 32(1) (subject to what has been said about third party claims) the contractor would be liable if his disregard of these contractual obligations caused damage to whatever had been found. The PM is specifically empowered by *clause 32(4)* to issue such instructions as he sees fit in relation to any object which is found. These instructions may be of any kind, but it is specifically stated that he has the right to instruct that a third party be allowed to examine, excavate or remove the object. The PM also has general power under clause 40(2)(m) to issue instructions about the action to be taken following the discovery of fossils, antiquities and objects of interest or value: see clauses 8 and 19 as to loss or damage to third parties etc. The operation of clauses 32(3) and (4) may give rise to two types of claim by the contractor. First, there may be a claim for extension of time under clause 36 on the ground that the finding of the object is a 'circumstance . . . which [is] outside the control of the contractor' or, in the alternative, because delay results from an instruction given by the PM. Second, any instruction of the PM will fall to be valued under clause 42 if it is a variation instruction or under clause 43 if it ranks as an ordinary instruction. In both cases the valuation will include an element for disruption etc. as appropriate.

Chapter 7

Commencement, programme, delays and completion

Introduction

This division of GC/Works/1 Edition 3 deals with the contractor's obligations as to progress and completion and with the inevitably related matter of delays to progress. In all lump sum contracts such as GC/Works/1 the contractor's obligation is to complete the defined work on or before the stipulated date. The only objective test of the contractor's performance under most standard contracts is the requirement of 'completion' itself.

Progress control is, in fact, at the heart of a problem in building which occurs far too frequently, i.e., late completion. The British Property Federation attempted to deal with the situation in its system for building design and construction, which was launched in 1983, by suggesting (*Manual of the BPF System*, 1983, p.53: 1983, British Property Federation, London) that all projects be monitored by means of project meetings, progress reports, and amendments to its schedule of activities.

These suggestions are given effect in Edition 3 of GC/Works/1, although different nomenclature is used. The new provisions do not, to the writer at least, appear to affect the basic principle that the employer or his representative should not interfere in the manner and timing of the execution of the works.

The requirement for a *programme* is a major new condition in Edition 3 and is coupled with provision for *progress meetings*. In the writer's view these provisions merely formalise what should be good practice in any event and a number of other modern contracts have quite rigorous requirements for the submission of a contractor's programme, e.g. the Singapore Institute of Architects Form, 1987 revision.

The effect of a main contract programme is often of more immediate importance to sub-contractors, as illustrated by *Martin Grant & Co. Ltd* v. *Sir Lindsay Parkinson & Co. Ltd* (1984). There sub-contractors agreed in 1971 to carry out their work 'at such time or times and in such manner as the contractor shall direct or require and observe and perform the terms of the [main contract] in so far as the same are applicable . . . .'. They also contracted to proceed with the sub-contract works 'expeditiously and punctually to the requirements of the Contractor . . . at such time or times as the Contractor shall require having regard to the requirements of the Contractor in reference to the progress or conditions of the Main Works . . . .'

The sub-contract contained no provisions corresponding to those in the main contract for extensions of time and reimbursement of prolongation and disruption costs, and in the event the sub-contractor found itself having to carry out the works several years after it had contemplated doing so. The sub-contractor was held to have no redress, because as Lord Justice Lawton said, the sub-contract provisions quoted above meant that:

'If the main contract was extended, then the sub-contractor's contract would be extended and that, during the time when the contract was in existence, the [sub-contractor] would do such portions of the work and at such time as might be required by the [main contractor]. In other words, there was a clear risk for the [sub-contractor] that the [main] contract and sub-contract might go on much longer than was originally contemplated.'

Commencement, programme, delays and completion

Clause 33 Programme

This is a major new provision and is important for a number of reasons. The contractor's progress obligations under clauses 31 and 34(1) are to proceed, *inter alia*, 'in accordance with the Programme' which is to be amended as necessary as work progresses: see, for example, clauses 35(3)(e) and 52(4)(a).

Clause 33(1) requires the contractor to produce a programme using 'the whole period for completion', and thus outlaws 'optimistic programmes': *Glenlion Construction Ltd* v. *The Guinness Trust* (1987). However, under clause 34(1) the contractor's obligation is to complete *'by the Date or Dates for Completion'* and so completion before the due date is clearly open to the contractor, but his programme must use the whole contract period. The programme must show the sequence of working, details of any temporary work, methods of work, labour and plant proposed and also events which in the contractor's opinion are critical to satisfactory completion of the works.

The programme will normally be required for submission with the tender and agreed with the Authority before the contract is let but *clause 33(2)* also allows for it to be submitted to the PM within 21 days after acceptance of the tender. Contrary to one ill-informed comment, the employer cannot impose a programme on the contractor.

The programming obligation is not a light one, however, since the programme must be fully resourced and is prepared at the contractor's cost. It is a matter for the contractor how he allocates the time for particular activities in excess of the time actually needed to carry out those activities, i.e., his 'float'. In practice this and other programming matters will be discussed and agreed with the PM, usually before the contract is made. The programme is an essential tool in evaluating the effects of unforeseen events and their impact on delivery dates. It will be of great assistance in regard to assessing extensions of time (clause 36) and determining claims for disruption and prolongation (clause 46). The programming obligation ties in with clause 35 dealing with *progress meetings*.

Clause 33(3) says that the programme is to be in the form and contain information relating to the manner and sequence of execution of the works as the PM may reasonably require. The PM will agree with the contractor the form of programme required, e.g., network analysis, precedence diagrams etc.

It is envisaged that any necessary amendments will be made to the programme following the monthly progress meetings under clause 35, and *clause 33(4)* emphasises that the contractor may submit proposals for amending the programme to the PM at any time. These proposals have no effect unless and until they are agreed by the PM.

Clause 34 Commencement and completion

Clause 34 is entirely different from the corresponding provision (clause 6) in Edition 2 and, unlike many standard form construction contracts no date for commencement is stipulated. In clause 1(1) 'the Date or Dates for Completion' are stated to mean 'the date or dates set out in, or ascertained in accordance with, the Abstract of Particulars', subject to any extensions of time awarded under clause 36. The abstract of particulars will also deal with the commencement date or the time periods within which possession of the site will be given to the contractor since it is not set out in the conditions. If a fixed completion date is stated in the abstract of particulars it is to be expected that a start date will also be specified there.

It should be noted that in any case it would be implied that possession of the site be given to the contractor within a reasonable time of acceptance of his tender (*Freeman* v. *Hensler* (1900)) and if, unusually, the contract were silent as to the contract period, the contractor's obligation would be to complete 'within a reasonable time' (*Startup* v. *Macdonald* (1843); *Charnock* v. *Liverpool Corporation* (1968)) This 'reasonable time' would be judged in the light of all the circumstances including the time which a reasonably diligent contractor would take: *Hydraulic Engineering Co. Ltd* v. *McHaffie, Goslett & Co. (1878)*.

Clause 34(1) first provides that 'within the period or periods specified in the Abstract of Particulars, the Authority shall notify the Contractor when he may take possession of the site or parts of the site'. This is a condition precedent to the contractor's obligation to commence. The Authority's notice to the contractor must be in writing: clause 1(3).

'The Site' is usefully defined in clause 1(1) as meaning 'the land or place which may be specified, allotted or agreed from time to time by the Authority as the place or places to be used for the purpose of carrying out the Contract'. The conditions themselves say nothing else about when possession is to be given, and so this must be dealt with in the abstract of particulars which ought also, desirably, to be more specific about the extent and boundaries of the site.

The degree of possession to be given depends on the circumstances of the case, but in principle in a project for new works the contractor is entitled to exclusive possession of the entire site, subject to the Authority's right (clause 65) to execute other works on the site contemporaneously: *London Borough of Hounslow* v. *Twickenham Garden Developments Ltd* (1970); *The Queen in Right of Canada* v. *Walter Cabott Construction Ltd* (1977).

This is a fundamental obligation imposed on the Authority and breach of it gives rise to a damages claim in respect of any loss suffered by the contractor: *Rapid Building Co. Ltd* v. *Ealing Family Housing Association Ltd* (1984). Under GC/Works/1, the Authority's failure to give possession as agreed gives rise to a claim for prolongation and disruption (clause 46(1)(b)) and extension of time (clause 36(2)(b)).

The legal position generally is put succinctly in *Hudson's Building and Engineering Contracts*, 10th edition, p.318 (1970, Sweet & Maxwell, London):

> 'Since a sufficient degree of possession of the site is clearly a necessary pre-condition of the contractor's performance of his obligations, there must be an implied term that the site will be handed over to the contractor within a reasonable time of signing the contract (see, e.g., *Roberts* v. *Bury Improvement Commissioners* (1870)) and, in most cases, it is submitted, a sufficient degree of uninterrupted and exclusive possession to permit the contractor to carry out his work unimpeded and in the manner of his choice.'

As stated, possession of the site is a precondition to the contractor's duty to commence the works and complete them by the specified completion date, as altered under the provisions relating to extensions of time or as a result of an acceleration agreement under clause 38. If the Authority fails or refuses to give possession of the entire site (as defined in clause 1(1)) then, as mentioned previously, that will ordinarily be a breach of contract entitling the contractor to damages, although in practice this breach will be dealt with under clause 36(2)(b) by the grant of an extension of time and by reimbursement under clause 46(1)(b) of the cost of any prolongation or disruption. However, it must be emphasised that the contractor has an alternative claim for damages at common law for failure to give possession of the site in accordance with the contract requirements.

Clause 34(1) goes on to provide that the date for completion is to be calculated from the date of possession so notified by the Authority. Failure by the contractor to complete by that date (subject to any award of extensions of time) is a breach of contract for which the remedy under the contract is liquidated damages.

If the contractor fails to complete the works and clear the site on or before the date for completion, the agreed liquidated damages are payable under clause 55 'for the period that the Works or any relevant section remain or remains uncompleted'. The amount of liquidated damages must be specified in the abstract of particulars and must, of course, represent a genuine pre-estimate of the likely loss caused by late completion or, as is more usual, a lesser amount. If no liquidated damages are specified then the Authority would need to rely on its right to unliquidated damages at common law subject to proof of loss.

Liquidated damages are recoverable without proof of loss under the simple deduction mechanism provided and in truth there is no reason for them not to be specified.

The Authority is given (clause 55(3)) an express right to deduct any liquidated damages from any monies due to the contractor. If the sum due as liquidated damages exceeds any further advance due to the contractor it is recoverable by the exercise of the Authority's contractual right of set-off from subsequent monies due or from any sum which may become due to the contractor under any other contract with the Authority or any other Government department: see clause 51.

Clause 55 is very tightly drafted; unlike JCT 80 a certificate of non-completion is not a pre-condition, nor is any notice from the Authority to the contractor a condition precedent, thus avoiding the problems of *A. Bell & Son (Paddington) Ltd* v. *CBF Residential Care & Housing Association Ltd* (1989), where the architect's failed to revise his completion certificate under a JCT 80 contract after granting further extensions of time and no fresh notice was given by the employer, so invalidating his entitlement to liquidated damages.

On the date specified in the Authority's notice the contractor is to take possession of the site and commence the execution of the works 'forthwith'. The contractor is then required to proceed with the execution of the works 'with diligence and in accordance with the Programme *or as may be instructed by the PM'* so that the whole of the works (or any relevant section) is completed to the satisfaction of the PM *by the date(s) for completion.* Presumably the italicised phrase refers to the powers of the PM to issue instructions under clause 40, e.g., clause 40(2)(e) under which the PM may give an instruction about the order of the execution of the whole or part of the works.

The wording indicates that substantial or practical completion is required, and the contractor is in breach only if he has not completed by the date specified in the abstract. Under clause 39, the PM must certify the date when the works (or any section) are completed to his satisfaction. He must issue a further certificate at the end of the longest relevant maintenance period specified in the abstract of particulars 'when the Works are in a satisfactory state'. These certificates are factual and are not made expressly dependent on the PM's opinion. Interestingly, clause

39(2) provides that any dispute about the giving of either of these certificates is to be referred to the Authority 'whose decision shall be final and conclusive', i.e., not subject to review in arbitration.

The contractor's obligation is to complete *by* the date for completion. He cannot be required to finish earlier than that date, unless he agrees to early possession of part under clause 37(1)(b) or to acceleration under clause 38.

Clause 34(2) is a subsidiary obligation. It requires the contractor to keep the site tidy and free from debris, litter and rubbish. No later than the completion date, the contractor must remove all unused materials from site and, by the due date, he must tidy up the site to the satisfaction of the PM. The contractor must comply at his own cost with any PM's instructions relating to the removal of goods, materials or rubbish.

Clause 35 Progress meetings

Clause 35 is one of the most important 'control' provisions in the contract, and expresses in formal language what should be good practice on any major project. The object of the clause is to provide regular opportunity for representatives of both parties to discuss *progress* and such matters as outstanding information required by the contractor, actual and potential causes of delay and the current situation as regards extensions of time.

Clause 35(1) provides for regular progress meetings which are normally monthly (clause 35(2)). It states that their purpose is 'to assess the progress of the Works .... and to facilitate their due and satisfactory completion' on time. The contractor's agent — who must be competent and in attendance on site during working hours: see clause 5 — must attend the meetings as the contractor's representative.

Clause 35(2) makes provision for these meetings to be held at monthly intervals, 'subject to any instruction to the contrary'. The PM specifies the time and place of the meetings.

Under *Clause 35(3)* three days before each meeting the contractor must send to the PM a comprehensive written report which must:

- describe progress by reference to the programme and any relevant PM's instructions
- specify outstanding information requests
- explain any circumstances which have delayed or might delay completion
- refer to any outstanding requests for extensions of time
- set out proposals for amending the programme to ensure completion by the due date

Clause 35(4) imposes a duty on the PM. Within seven days of each progress meeting he must give the contractor a written statement specifying:

- the extent to which he considers the project is on time, delayed or early
- matters which he considers have delayed or are likely to delay completion
- the steps he has agreed with the contractor to reduce or eliminate any delay
- the situation about extensions of time

- his response to outstanding requests for drawings, nominations and other information.

Following each progress meeting, the PM is to amend the stage payment charts if he 'has recorded in a statement after a progress meeting that the Works are in delay or ahead of Programme' and the contractor's entitlement to payment is adjusted accordingly: see clause 48(3).

Clause 36 Extensions of time

Extensions of time are dealt with by clause 36 which is a radically re-written version of its predecessor and is to be regarded as one of the better standard form extension of time clauses. A point to notice is that there is no right to an extension of time for weather conditions which are, therefore, at the contractor's risk. The scheme of the clause is such that in the majority of cases the PM will give consideration to the grant of an extension of time after receipt of a notice from the contractor, but the clause recognises explicitly that the PM can act of his own volition. Indeed, it is essential that he should do so if he is aware that delay has been or is likely to be caused by any act or default of the Authority, or those acting on its behalf, which is going to cause delay to completion.

If the PM fails to grant an extension of time in those circumstances, time will become 'at large' under the contract. The contractor's obligation would then be to complete within a reasonable time: *Fisher* v. *Ford* (1840). The consequence would be that there is no date from which liquidated damages (clause 55) could begin to run and so the Authority would lose its right to liquidated damages since if any part of the delay is caused by the Authority, no matter how slight, and no extension of time is grantable under the contract and actually granted, the liquidated damages clause becomes inoperative: see, for example, the well-known case of *Peak Construction (Liverpool) Ltd* v. *McKinney Foundations Ltd* (1970).

Clause 36(1) is the provision under which any alteration to the completion date is made. It modifies the contractor's liability to complete the works by that date (clause 34(1)) and to pay liquidated damages for the breach of late completion (clause 55). Although the operation of clause 38 (acceleration) and clause 52 (cost savings) may require an adjustment of the specified completion date, it is under this sub-clause that the alteration is made. Clause 36(1) requires the PM to grant an extension of time in specified circumstances and, while clause 46 provides for the contractor to be reimbursed for the financial effects of both prolongation and disruption, there is no necessary connection between the grant of an extension of time and a claim for financial reimbursement.

Clause 36(1) requires the PM to grant an extension of time where he 'receives notice requesting an extension of time from the contractor or where he considers that there has been or is likely to be a delay which will prevent completion ... by the relevant Date for Completion'. This recognises explicitly that extensions of time can be issued in the absence of a request from the contractor.

The PM must notify the contractor of his decision 'as soon as possible

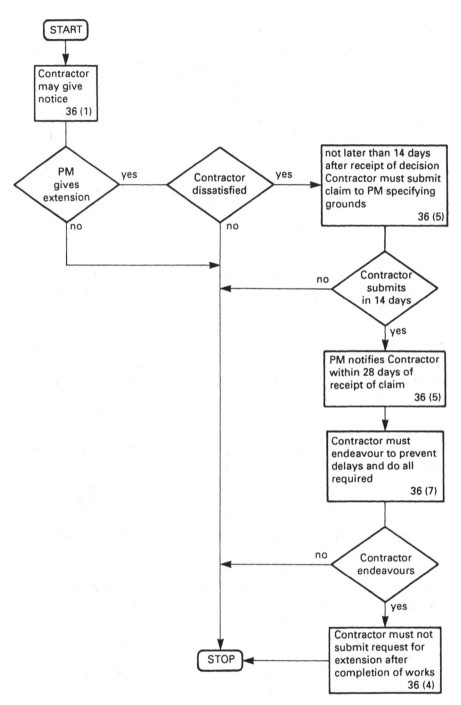

Fig. 7.1: Extensions of time: Contractor's duties

Commencement, programme, delays and completion 113

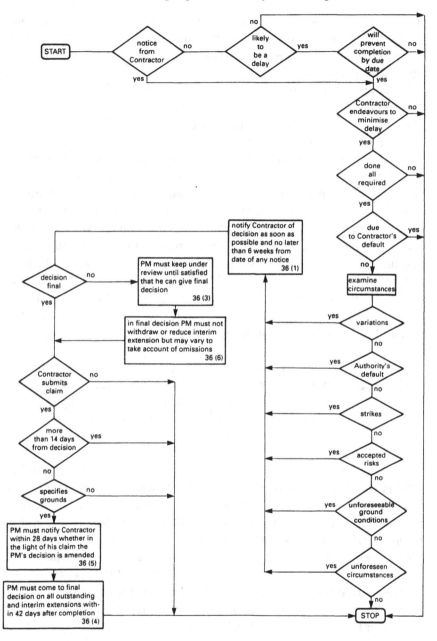

Fig. 7.2: Extensions of time: Project Manager's duties

and in any event within 6 weeks from the date of any notice he has received' from the contractor. This time limit — and the other time limits imposed by the clause — are sacrosanct and cannot be extended by agreement: see clause 1(3).

Nothing is said about the content of the contractor's notice, but it is suggested that it must be specific in its terms. The contractor must give written notice to the PM as soon as he becomes aware that there *has been* or *is likely to be* delay which will prevent completion of the works or any section by the relevant completion date. The notice should give details of the circumstances which have caused or are likely to cause delay and, desirably, an estimate of the period of extension required. The notice should also identify one or more of the specified causes. Such matters will be discussed at the monthly progress meetings under clause 35. If the contractor fails to give notice, this is a breach of contract and there is authority for the view that such a breach can be taken into account by the PM when granting an extension of time: *London Borough of Merton* v. *Stanley Hugh Leach Ltd* (1985). In that case, Mr Justice Vinelott approved of a statement in Keating's *Building Contracts*, 4th edition (1970, Sweet & Maxwell, London) to the following effect:

'[The architect can] take into account that the contractor was in breach of contract and must not benefit from his breach by receiving a greater extension than he would have received had the architect, upon notice at the proper time, been able to avoid or reduce the delay by some instruction or reasonable requirement.'

Clearly the PM must consider all the circumstances in deciding whether or not to grant an extension of time, and he must have regard to the actual situation since the effect of the delaying event is to be assessed at the time the work is actually being carried out : *Walter Lawrence & Son Ltd* v. *Commercial Union Properties (UK) Ltd* (1984). The PM is entitled (clause 36(6)) to take the effect of any omission instructions into account.

Clause 36(2) sets out the grounds — and they are the only grounds — on which an extension of time may be awarded for the delay or likely delay. There are only six events specified as giving rise to a potential entitlement to extension of time. They are:

- the execution of any modified or additional work
- the act, neglect or default of the Authority or the PM
- any strike or industrial action which prevents or delays the execution of the works and which is outside the control of the contractor or

any of his sub-contractors (including nominated sub-contractors)
- an accepted risk, i.e., insurance risks, which is given a limited definition in clause 1(1)
- unforeseeable ground conditions notified to the PM under clause 7(3); they are defined there as 'ground conditions (excluding those caused by weather but including artificial obstructions) which [the contractor] did not know of, and which he could not reasonably have foreseen having regard to any information which he had or ought reasonably to have ascertained'
- any other circumstances (other than weather conditions) which are outside the control of the contractor or any of his sub-contractors and which could not have been reasonably contemplated under the contract. This broad 'sweeper' must be given a restrictive interpretation and in essence appears to be aimed at what are usually called Acts of God: see *Lebeaupin* v. *Crispin* (1920) discussing the similar (if wider) concept of *force majeure*.

Clause 36(3) is procedural. The PM is bound to say whether his decision is interim or final. Interim decisions must be kept under review 'until [the PM] is satisfied from the information available to him that he can give a final decision'. It is important that any interim decision should be realistic and take into account the facts reliably known to the PM at the time.

Clause 36(4) obliges the PM to reach a final decision on all outstanding and interim extensions of time within 42 days after completion of the works, i.e., 42 days after the date of completion certified by him under clause 39(1). Completion of the works is a cut-off point since no requests for extensions can be submitted by the contractor thereafter. In his final decision, the PM is not entitled to withdraw or reduce any interim extensions already awarded.

The clause does not require the PM to give reasons for his award (or refusal of an award) but it is suggested that he should do so, preferably by reference to the appropriate sub-paragraphs of clause 36(2) and the programme. It is also important that the award should always specify clearly the revised completion date.

The granting of extensions of time is not an exact science and there are acute problems in the case of overlapping and cumulative delays. Reference should be made to *H. Fairweather & Co. Ltd* v. *London Borough of Wandsworth* (1988) which discusses some of the practical problems involved and disapproves of the common practice of granting an extension of time for the 'dominant' reason for the delay. There is no mechanism under GC/Works/1 for allocating an extension of time between the different heads of award listed in clause 36(2).

Clause 36(5) says that if the contractor is dissatisfied with the PM's decision under clause 36(1), he has 14 days from its receipt in which to submit a claim to the PM specifying the grounds which in his view entitle him to an extension or further extension of time. The PM has 28 days from receipt of the contractor's claim in which to reconsider the matter and notify the contractor 'whether in light of his claim the disputed decision has been amended'.

Clause 36(6) says:

'In considering any extension of time the PM may vary any interim decision .... to take account of any authorised omission from the Works or relevant Section.'

Obviously, the PM will take account of authorised omissions which have the effect of saving time. This sub-clause makes it clear that he can vary an interim decision in light of authorised instructions made subsequent to the award. It ties in with his obligation (clause 36(3)) to keep interim decisions under review.

Clause 36(7) governs the whole of the clause. The contractor loses his entitlement to an extension of time where 'the delay or likely delay is, or would be attributable to, [his] negligence, default, improper conduct or lack of endeavour'. This restriction is self-explanatory. The sub-clause also imposes a positive duty on the contractor. He is obliged to 'endeavour to prevent delays and minimise unavoidable delays, and to do all that may be required to proceed with the Works'. This is an obligation which is more clearly expressed than its JCT 80 equivalent, and is one which certainly does not require him to expend substantial sums of money.

Flow chart Figs 7.1 and 7.2 and illustrate the duties of the contractor and the PM in relation to extensions of time.

Clause 37 Early possession

This clause deals with the situation where the Authority wishes to take possession of part of the works before completion of the whole. It applies both where the works are to be completed in phased sections *and* where the Authority decides during the course of the contract that it would like to take possession of part of the works. It follows closely condition 28A of Edition 2.

Clause 37(1) says that the Authority shall be entitled to take possession of any part of the works which the PM has certified under clause 39(1) is completed to his satisfaction provided:

(1) the completed part is a section, i.e., is specified as such in the abstract of particulars, and which has a specified date for completion (clause 1(1)); *or*
(2) the completed part is a part of the works (including part of a section) in respect of which the parties *agree* that possession shall be given before completion of the works as a whole.

The 'completed part' is the part of the works of which possession is taken by the Authority and which is certified by the PM as having been completed to his satisfaction under clause 39(1).

On and after the date on which the PM certifies his satisfaction with the completed part it 'shall no longer form part of the Works for the purposes of' clause 19 (loss or damage), clause 30 (vesting) and clause 8 (insurance). The contractor is thus relieved of responsibility in this regard in respect of the part of the works handed over to the Authority.

Clause 37(2) provides that the maintenance period for the purposes of clause 21 (defects) runs from the date of the PM's certificate under clause 37(1) 'in respect of the completed part or any sub-contract works comprised in it'. The obligations in clause 21 apply to the completed part from the date of the PM's certificate.

Clause 37(3) says that, *as soon as possible* after the date of the PM's clause 37(1) certificate, the PM must certify the value of the completed part for the purposes of liquidated damages and payment of the reserve (see clause 48) to the contractor.

The operation of *clause 37(4)* prevents the happening of the sort of problems which occurred in *Bramall & Ogden Ltd* v. *Sheffield City Council* (1983) where a local authority employer was held disentitled to liquidated damages because there was no mechanism in the contract for dealing with the sectional completion which occurred. It deals effectively with the effect of early possession on the calculation of liquidated damages. It provides that the liquidated damages provision (clause 55) remains in force, but the contractor's liability is to be reduced: where the completed part comprises part of the works but not a section or sections, the rate of liquidated damages is reduced by the same ratio as the value of the completed part bears to the value of the works or relevant section of them.

The value of the completed part and the work or any relevant section is to be determined by the PM.

Clauses 37(5) and 37(6) deal with payment. The reserve (retention) accumulated by the Authority under condition 48 is to be apportioned by the PM as at the date of his certificate under clause 37(1). The apportionment is to be made 'so that the share of the reserve apportioned in respect of a completed part shall bear the same ratio to the whole of the reserve as the value of the completed part bears to the value of the Works' as estimated by the PM at that date.

Under *clause 37(6)* the Authority must then pay to the contractor half of the share so apportioned in respect of the completed part, i.e., the first half of the reserve is paid over immediately: paragraph (a). The remaining half is to be paid to the contractor after the end of the maintenance period for the completed part, when the PM has certified that the completed part is in a satisfactory state under clause 39(1): paragraph (b).

Clause 37(6) makes any decision of the PM under clause 37 'final and conclusive' thus removing it from review in arbitration or otherwise.

Commencement, programme, delays and completion 119

Clause 38 Acceleration

Clause 38 provides the framework if the Authority wishes the works or any section of them to be completed before the specified date(s) for completion. Its operation depends on the agreement of the parties. If the contractor agrees to accelerate and his proposals are acceptable to the Authority, the detailed consequences on price and working methods must be agreed.

Clause 38(1) provides that if the Authority wants the works or a section completed before the due date, it can direct the contractor to submit priced proposals for acceleration and any consequential amendments to the programme; alternatively it provides that he submit his 'explanation why he is unable to achieve the accelerated completion date'. This must be done within a period specified in the Authority's direction.

Clause 38(2) provides for what is to happen if the Authority accepts the contractor's proposals. The Authority must then specify in writing (see clause 1(3)):

- the accelerated date for completion of the works or relevant section
- the amendments to the programme (clause 33), including any relevant critical paths and supporting documentation
- the amount by which the contract sum is to be increased
- revised stage payment charts (clause 48)
- any other relevant agreed amendments to the contract.

Under *clause 38(3)* the contractor can submit proposals for accelerated completion of his own volition and the Authority 'undertakes to consider' them. If the contractor submits proposals voluntarily and these are accepted by the Authority, then clause 38(2) will be applicable.

Clause 39 Certifying work

This straightforward provision is a simplified version of clause 42 in Edition 2. It deals with the issue of certificates in respect of work and says something about their nature and effect. Payment certificates are, sensibly, dealt with under a separate provision (clause 50).

Clause 39(1) imposes a duty on the PM to certify the date when the works (or any section) are completed to his satisfaction. This triggers off the operation of clause 49 (final account). At the end of the longest maintenance period specified in the abstract of particulars, the PM is to issue a further certificate when the works are in a satisfactory state. This enables final payment to be made to the contractor (see clause 49(4)(5)) and the contractor's insurance liability ends (see clause 8(1)).

The consequences of the completion certificate under clause 39(1) are:

- the contractor's liability for liquidated damages ends (clause 55)
- the contractor is entitled to be paid an amount which equals the difference between the PM's estimate of the final sum, less one half of the reserve accumulated, and the total advances already paid (clause 49(1))
- the maintenance period begins to run (clauses 21 and 49)
- any difference or dispute can be referred to arbitration (clause 69).

Clause 39(2) provides that any dispute as to the contractor's right to a certificate is to be referred at his request to the Authority whose decision is made final and conclusive.

Commencement, programme, delays and completion

Table 7.1: Contractor's powers under GC/Works/1 — Edition 3

Clause	Power	Comments
33(4)	Submit proposals to the PM for amending the programme.	At any time, without prejudice to clause 35. PM must agree before the proposals are included in the programme.
38(3)	Submit proposals to the authority for completing before completion date or dates.	At any time. The Authority must consider them and, if acceptable, take clause 38(2) action.
45(5)	Refer the matter to HM Commissioners of Customs and Excise.	If the contractor is not satisfied that the Authority has correctly allowed for tax in any payment made.
52(1)	Submit a written proposal to the PM which the contractor thinks will reduce the cost of the works or maintenance or increase the efficiency of the completed works.	At any time during the contract. The proposal must state it is submitted under this clause and include an estimate of the contractor's share of any savings.
59(1)	Refer any dispute etc. arising out of the contract for adjudication to person named in abstract of particulars.	If not resolved within three months and not a matter on which a decision is final and conclusive.
63(6)	Make reasonable objection to a person nominated as sub-contractor or supplier.	Contractor must provide such information as the PM reasonably requires in relation to any objection.

Chapter 8

Instructions and payment

Introduction

Clauses 40 to 52 deal with instructions and financial matters. They introduce significant changes as compared with the corresponding provisions of Edition 2.

Amongst the PSA's objectives in producing Edition 3 was the need to speed up the agreement of final accounts and to contain contractors' claims for prolongation and disruption within bounds, as well as to deal with the controversial issue of finance charges as a head of claim.

Following the decision of the Court of Appeal in *F.G.Minter Ltd* v. *Welsh Health Technical Services Organisation* (1980), which was later amplified by the decision of the same court in *Rees & Kirby Ltd* v. *Swansea City Council* (1985), the PSA sought counsel's advice as to the meaning of the word 'expense' used in condition 9(2) of Editions 1 and 2, and in condition 53 of Edition 2. Counsel advised that the term included finance charges in the *Minter* sense, i.e., both interest paid on money borrowed and interest not earned where the contractor deployed his own capital in financing the project, as being a constituent part of the 'expense' incurred, in contrast to interest for late payment which is not recoverable at common law: *London Chatham & Dover Railway Co.* v. *S.E.Railway Co.* (1893); *President of India* v. *La Pintada Cia. Navegaçion SA* (1984). Despite or perhaps because of these decisions there has been continuing uncertainty as in some circumstances interest may be recoverable as a head of special damages: *President of India* v. *Lips Maritime Corporation* (1987); *Holbeach Plant Hire Co. Ltd* v. *Anglian Water Authority* (1988). Edition 3 addresses all these problem areas.

The PSA was also attracted by the British Property Federation system

Instructions and payment

of lump sum agreement for variations, introduced in both the ACA Form of Building Agreement 1982 (2nd edition 1984) and in the ACA/BPF Form of Building Agreement 1984 and taken aboard by the Joint Contracts Tribunal, somewhat reluctantly, in its supplementary provisions to the JCT Standard Form of Building Contract with Contractor's Design 1981 published in 1988. A more workable variant of this concept is therefore introduced by clauses 40(6) and 42.

It was also the wish to introduce a new scheme of interim payments which would act not only as an incentive to contractors, but also make for administrative simplicity. Clause 48 accordingly introduces the use of stage payments fixed in accordance with a stage payment chart. This means that, provided the contractor achieves the necessary rate of progress, he knows in advance the exact amount he will receive each month, apart from the value of variations and any payments for disruption or prolongation.

Variations are well-known to be the commonest cause of disruption and prolongation claims and the PSA working party grasped the mettle by excluding them from clause 46 (which deals with prolongation and disruption claims) and which is now limited to what are effectively 'breach of contract' type claims. The valuation provisions for variation instructions now provide for the valuation to include for any cost of disruption to or prolongation of both varied and unvaried work consequential on compliance with the variation instruction.

Edition 3 also tightens up procedures for the settlement of the final account in order to forestall protracted argument over the final sum. In the PSA's experience (as well as in that of many employers under other standard forms of building contract) the financial completion of projects was taking an excessive time. Indeed, in some cases known to the writer, it takes longer to prepare and agree the final account than it does to construct the building! Clause 49 deals with this problem in an effective way.

Clause 40 PM's instructions

Clause 40 is a redrafted and substantially revised version of clause 7 in Edition 2. It sets out with precision the matters on which the PM is empowered to issue instructions. If an instruction is empowered by the contract, the contractor must comply with it and his failure to do so is a breach of contract. If he fails to comply with an instruction, clause 53 empowers the PM to issue a notice requiring compliance, and if the contractor still fails to comply within the period specified in the notice the Authority is empowered to have the work done by others and claim reimbursement of its costs and expenses from the defaulting contractor. This power in clause 53 is 'without prejudice to the exercise of [the Authority's] powers to determine the contract' under clause 56, which is the ultimate sanction.

Clause 40 also deals with the confirmation of the few oral instructions which the PM is empowered to give, and emphasises (clause 40(4)) that the contractor is not to vary the works as described or shown in the specification, bills of quantities and drawings except in accordance with an instruction from the PM.

The clause now distinguishes between a *variation instruction* (VI) which is defined as one altering, adding to or omitting something from the works or requiring any change in their design, quality or quantity, and instructions dealing with secondary matters. In the former case, the PM can require the contractor to submit a lump sum quotation for the variation.

Clause 40(1) empowers the PM to issue 'from time to time further drawings, details, instructions, directions and explanations'. Although the sub-clause is phrased in permissive terms, it is quite clear under the general law that, as agent of the Authority, the PM must issue plans, drawings and other information necessary for the execution of the works and at the proper time, i.e., within a reasonable time. This means that the information must be issued as the contract progresses and at times which are reasonable in all the circumstances having regard to the contractor's progress and the programme (clause 33) which will be amended regularly. Reference should be made to clause 35, dealing with progress meetings: clause 33(3)(b) provides that one of the items in the contractor's pre-meeting written report is to 'specify all outstanding requests .... for drawings, nominations, levels or other information', while the PM's post-meeting written statement is to set out his response: clause 33(4)(e).

The well-known case of *Neodox Ltd* v. *Borough of Swinton and Pendlebury* (1958) discusses the common law position and also what is a 'reasonable time' in the context of a civil engineering contract, and it is submitted that the position is the same under GC/Works/1. Failure by the PM to issue the necessary design and other information timeously is a breach of contract for which the Authority is liable and, indeed, under clause 46 the contractor may have a claim for prolongation and disruption expenses in respect of late information supplied by the PM (see clause 46(2)(c)).

However, GC/Works/1 does not exclude claims at common law and so the contractor has an alternative claim for breach of contract if necessary information is supplied late and delay is caused thereby: *London Borough of Merton* v. *Stanley Hugh Leach Ltd* (1985).

Moreover, it is submitted that it is an implied term of the contract that the Authority will procure that the PM issues the drawings, details, instructions, directions and explanations that are necessary to enable the contractor to fulfil his obligations.

The position was fully discussed in *Perini Corporation* v. *Commonwealth of Australia* (1969) where it was held that the Government was bound to ensure that its Director of Works performed the various duties imposed upon him by a building contract. This ruling is of equal application to GC/Works/1.

Clause 40(2) particularises the instructions which the PM is empowered to give. He may issue instructions about any of the following matters:

- ***The variation or modification of all or any of the specification, drawings or bills of quantities, or the design, quality or quantity of the works.***

This statement in fact defines 'variation' (see clause 40(5)) and thus also sets the limits on the PM's power to order variations. The PM's power is to order variations to the work as originally described in the bills and/or specification and/or drawings and there are many reported cases in which changes ordered have been held to fall outside particular variation clauses, many of them of Transatlantic origin: *Watson* v. *Bennett* (1860); *Reid* v. *Batte* (1829); and *Boyd* v. *South Winnipeg* (1917) are examples. Reference may usefully be made to *Variations in Construction Contracts* by Peter R. Hibberd (1986, Collins Publishers, 1986) chapter 1. The important point is that any variation ordered must have some direct relationship to 'the Works' as defined. If it does not, then it goes to the root of the contract.

Variations ordered under clause 40(2) fall to be valued under clause 42 and the execution of any modified or additional work may give rise to an extension of time under clause 36(2)(a).

- *Any discrepancy in or between the specification, drawings and bills of quantities*

It is thought that such an instruction might constitute a variation, depending on the context. The provision allows the correction of errors and the contractor is bound, under clause 2(3), to inform the PM of any discrepancies which he notices when he is handling the contract documents. It is a matter for the PM's discretion as to whether to issue instructions under this provision to deal with discrepancies or inconsistencies or to issue an instruction requiring a variation.

- *The removal from site of any things for incorporation and their substitution with other things.*

If such an instruction is given because the items are rejected by the PM under clause 31(4) it would not rank for payment because of the provisions of clause 31(7(b).

- *The removal and/or re-execution of any work executed by the contractor.*

This must be read in conjunction with clause 21 relating to defects during the maintenance period and to clause 31(7) covering the contractor's liability to replace, rectify or reconstruct condemned work at his own expense.

- *The order of the execution of the whole or part of the works.*

- *The hours of working and the extent of overtime or night work to be adopted.*

These matters may well be dealt with in the contract documents or in the contractor's programme submitted under clause 33. Instructions of this nature may seriously affect the basis on which the contractor has tendered, and it is to be noted that (unlike the analogous provision in clause 14 of JCT 80) the contractor has no right of reasonable objection to compliance with such an instruction. Equally, an instruction under this paragraph does not fall within the definition of 'variation'.

However, if the PM issues an instruction under clause 40(2)(f) it will fall to be valued under clause 43 and the valuation will take account of the effect of any disruption or prolongation.

- *The suspension of the execution of the whole or any part of the works.*

At common law the employer has no power to direct suspension of the works under a construction contract in the absence of an express term empowering him to do so: see *Determination and Suspension of Construction Contracts* by Vincent Powell-Smith and John Sims, Chapter 7 (1985, Collins Publishers, London).

- *The replacement of any person employed in connection with the contract.*

See clause 6 as to the PM's power to require removal and replacement of the contractor's employees.

- *The opening up for inspection of any work covered up.*

Clause 31(4) deals with the PM's power of inspection and testing, while clause 17 requires the contractor to give due notice before covering work and adds that 'in default of notice the Contractor shall, if required by the PM, uncover the work or Things at his own expense'.

The position would, therefore, appear to be that if the inspection was in the contractor's favour, the instruction would attract payment under clause 43, except if it resulted from the contractor's failure to give notice under clause 17.

- *Amending and making good of defects.*

This cannot normally qualify for payment since an instruction under this paragraph will result from the contractor's default. The provisions of clause 21 as to defects appearing during the maintenance period should, however, be noted since, after completion of the remedial works under that clause, the contractor may be able to establish a claim to payment.

- *The execution of any emergency work.*

Clause 54 obliges the contractor to carry out any emergency work (as defined in clause 54(3)) required by the PM and, unless that emergency work resulted from the contractor's own default, the instruction would fall to be valued under clause 43.

- *The use or disposal of material obtained from site excavations.*

Under clause 32(2) the Authority may permit its property (including excavated materials) to be used in substitution for any things for incorporation. In that event the sub-clause provides for the contract sum to be reduced by the amount of any saving as ascertained by the quantity surveyor. If the PM directs disposal of excavated materials under this paragraph, in some cases the instruction might rank for payment under clause 43, e.g., if the use instructed was not reasonably foreseeable by a competent and experienced contractor.

- *Action to be taken following discovery of fossils, antiquities and objects of interest and value.*

Clause 32(3) deals with the finding of such items and clause 32(4) empowers the PM to issue the necessary instructions. These would rank for payment under clause 43.

- *Measures necessary to avoid nuisance or pollution.*

Under clause 14 the contractor is obliged to take reasonable precautions to avoid nuisance or pollution, and it is clear that the cost of compliance is on the contractor. An instruction issued under this paragraph would not, therefore, in principle, rank for any payment.

- *Any other matter which the PM considers necessary or expedient.*

By the terms of *clause 40(3)* the decision of the PM as to whether any necessary instruction is necessary or expedient is made final and conclusive. In practice, the limitation on the apparently unfettered discretion conferred on the PM is that the instruction must refer to some aspect of the works, progress etc. The instruction must bear some relationship to the subject-matter of the contract and compliance by the contractor is subject to legal and physical impossibility.

Conceivably, the provision might be used to dictate the contractor's methods of working, although despite the apparently unqualified words it must be remembered that 'it is the function and right of the builder to carry out his operations as he thinks fit': *Clayton v. Woodman & Son Ltd* (1962).

Whether or not an instruction under this paragraph ranks for payment under clause 43 will, of course, depend on its type.

Clause 40(3) provides that all instructions must be in writing except those:

- correcting discrepancies in or between the contract documnts
- requiring removal or re-execution of any work executed
- suspending the execution of the whole or part of the works
- requiring the execution of emergency work.

Such instructions may be given orally but must be confirmed in writing within seven days and, while writing is not expressly made a condition precedent to entitlement to payment under clauses 42 and 43, it is essential that there be written confirmation. If the PM does not confirm an oral instruction within seven days, it is desirable that the contractor should write seeking his written confirmation and that the time limit be agreed to be waived: see clause 1(4).

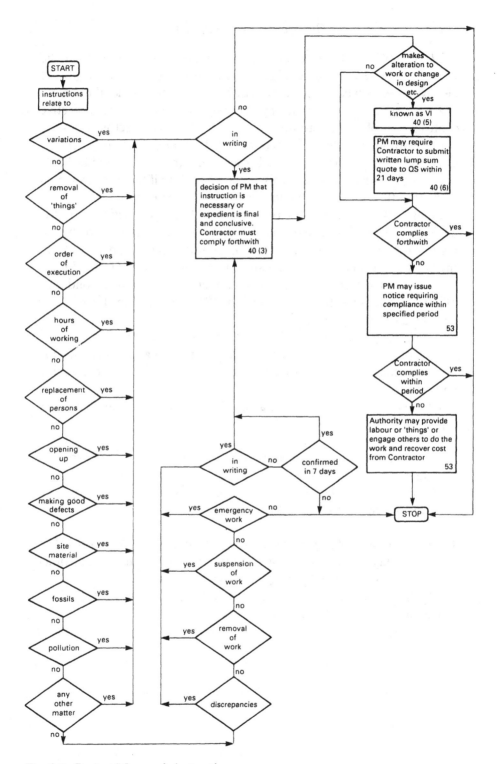

Fig. 8.1: Project Manager's instructions

Clause 40(1) prohibits the contractor making any variation in the works otherwise than in accordance with an instruction of the PM. It merely expresses what would otherwise be the situation under the general law. There is no provision in the contract for the retrospective confirmation of an unauthorised variation. If the contractor varies the works without authority he has no right to any extra payment or extension of time and he is also in breach of contract and would be liable to at least nominal damages. This would be so even if the work as varied without authorisation was better than that contracted for.

Clause 40(5) distinguishes two types of PM's instructions for the purposes of valuation. The first is a variation instruction (VI), which is an instruction which makes 'any alteration or addition to or omission from the Works or any change in the design, quality or quantity of the Works'. Variation instructions fall to be valued under clause 42. The second type of instruction, by implication, is any other PM's instruction. These fall to be valued under clause 43.

Clause 40(6) empowers the PM to include in any VI a requirement for the contractor to submit a written lump sum quotation 'showing the amount which the contractor estimates to be the *full cost* of complying with' the VI.

Clause 42(2) requires the contractor's lump sum quotation must show seperately:

- The direct cost of complying with the instruction, i.e., the variation must be priced; and,
- any consequential cost of disruption to or prolongation of both varied and unvaried work.

If the VI requires a lump sum quotation, the contractor must provide it within 21 days. The contractor's quotation must include 'such other information as will enable the QS to evaluate' it: clause 42(2).

The flowchart Fig. 8.1 illustrates the operation of the PM's powers while Table 8.1 tabulates them.

Instructions and payment

Table 8.1: Project Manager's powers under GC/Works/1 — Edition 3

Clause	Power	Comments
2(2)	Instruct that the specification shall not take precedence over the drawings.	
2(5)	Decide suitability of form of contract drawings, specification, blank bills of quantities and drawings issued during the progress of the works.	
	Have access to drawings and specification on site.	At all reasonable times
4(2)	Expressly delegate powers and duties to named representatives.	Notice must be given to the contractor. The appointment of representatives does not prevent the PM from exercising any of the powers under the contract.
5	Require the contractor's agent to attend at the office of the PM.	At any time.
6(1)	Require the contractor immediately to cease to employ any person in connection with the contract.	If the PM thinks his continued employment is undesirable. The contractor must replace with suitably qualified person. PM's decision is final and conclusive.
7(7)	Agree that ground conditions in notice under 7(3) could not reasonably have been foreseen by the contractor having regard to information he should have had in accordance with 7(1) and 7(3).	The PM must certify them unforeseeable ground conditions and notify the contractor.
10(1)	Require the contractor to submit for approval a suitable drawing or other design document relating to that work. Approve the design in writing.	If the contract requires the contractor to design any part of the works. Before the contractor commences work.
	Approve alteration of the design.	None of these approvals relieves the contractor of design liability.
15	Direct the contractor's agent as to the form of daily labour and plant return.	
16	Approve excavations for foundations.	
17	Require the contractor to uncover anything covered up.	In default of notice.
25(2)	Have access to records required under clause 25(1)	
26(1)	Give notice to the contractor that any person is not to be admitted to site.	Contractor must take all reasonable steps to prevent admission.

Table 8.1: Project Manager's powers under GC/Works/1 — Edition 3 (cont.)

Clause	Power	Comments
	Reasonably require the contractor to take steps to prevent unauthorised admissions to site.	
26(2)	Instruct the contractor to give list of names and addresses of all persons who are or may be concerned with the works.	Contractor must specify capacities in which they are so concerned.
26(3)	Make final and conclusive decision regarding whether: • A person is to be admitted to site. • Contractor has furnished information or taken steps required by this clause.	
27	Reasonably require information from the contractor. Demand that the passes be returned.	In order to issue passes. At any time.
28	Consent to the taking of photographs of the works, publication or circulation of the same.	
30(4)	Consent to 'things' being removed from site. Give the contractor written instruction or permission to remove from site unused or rejected 'things'.	Consent must be in writing. At any time during the execution of the works.
31(4)	Request the contractor to show that he is performing his duties under clauses 31(1) and 31(2). Inspect and examine any part of the works or inspect, examine and test any 'thing for incorporation' on site in any factory etc. Reject 'things' which do not conform to specification, bills of quantities or drawings or which are not of good quality or fit for purpose.	 Contractor must give reasonably required help and facilities. In the PM's opinion.
31(5)	Arrange for an independent expert to test whether 'thing for incorporation' is fit for use in the works, of good quality and conforms to contract.	Contractor must bear cost if result shows 'thing' is not substantially in accordance with the contract. If satisfactory but negative test results, he must bear costs of more tests required to monitor quality. The expert's report is final and conclusive.
32(4)	Issue instructions in relation to any object falling under clause 32(3).	As he sees fit. Contractor may be required to permit removal etc. by a third party.
33(2)	Agree the contractor's draft programme.	Not more than 21 days from acceptance of tender unless submitted before acceptance of tender and agreed by authority.

Instructions and payment

Clause	Power	Comments
33(3)	Reasonably require the programme to be in particular form and contain information on manner and sequence of execution of works.	
33(4)	Agree contractor's proposals for amending the programme.	
34(1)	Instruct the contractor regarding the progression of the works.	
36(6)	Vary any interim decision to take account of authorised omissions from the works or any section.	In considering any extension of time.
37(1)(b)	Instruct that possession is to be given before completion of the works.	Decision is to be final and conclusive.
40(1)	Issue further drawings, details, instructions, directions and explanations.	From time to time. All are to be treated as instructions.
40(2)	Issue instructions in relation to: • Variations. • Discrepancies. • Removal of 'things for incorporation' and substitution of other 'things'. • Removal or re-execution of work. • Order of execution of works. • Hours of working and extent of overtime. • Suspension of works. • Replacement of any employee. • Opening up for inspection. • Making good defects under clause 21. • Emergency work under clause 54. • Use of excavated material. • Actions after discovery of fossils, etc. • Measures to avoid nuisance and pollution. • Any other matter which the PM considers necessary or expedient.	The PM's decision that any instruction is necessary or expedient is to be final and conclusive. All instructions must be in writing except with regard to discrepancies, removal or re-execution of work, suspension of work or emergency work which must be confirmed within seven days. Contractor must comply with instructions forthwith.
40(6)	Include a requirement for the contractor to submit to QS lump sum quotation for cost of compliance.	In a VI.
42(1)(a)	Accept a lump sum quotation prepared by the contractor and submitted to the QS in accordance with clauses 42(2) and 42(3).	In order to determine the value of a VI. The lump sum must show how it was calculated and

Table 8.1: Project Manager's powers under GC/Works/1 — Edition 3 (cont.)

Clause	Power	Comments
		the contractor may include other information to help the QS evaluate the quotation.
48(6)	Request the contractor to show that any amount or supplier of 'things for incorporation' covered by a previous advance has been paid.	Before payment of any advance or the issue of the final certificate. The Authority may withhold payment if the PM is not satisfied.
52(2)	Require the contractor to provide more information.	In respect of the contractor's cost saving proposals.
54(1)	Require the contractor to carry out emergency work.	
61(4)	Submit representations to the adjudicator.	Not later than 14 days after receipt of a contractor's notice of reference to adjudication in accordance with clause 61(1).
62(1)	Give consent to sub-letting. Require the contractor to provide full details of a proposed sub-contractor.	
63(2)	Instruct the way in which a person may be nominated or appointed. Consent to the contractor ordering work or 'things' under prime cost items.	Prior to the conclusion of an authorised sub-contract.
63(6)	Require the contractor to provide information.	In relation to an objection which the contractor may make to a prospective nominated sub-contractor or supplier under clause 63(6).
64	Instruct work to be carried out under provisional lump sums and provisional items.	To be ascertained in accordance with clause 42.

Clause 41 Valuation of instructions — principles

Clause 41 deals with the principles involved in the valuation of instructions and recognises the basic division between VIs (clause 40(5)) and all other instructions. It is entirely new. Clause 41(1) draws the distinction. In both cases the valuation must include the cost (if any) of any disruption to or prolongation of both varied and unvaried work: *clause 41(2)*.

Clause 41(3) is self-explanatory. It states that the value of any instruction shall be added to or deducted from the contract sum, except that there is no financial adjustment where the instruction was necessary because of any default or neglect by the contractor or any of his servants, agents or sub-contractor, *whether nominated or domestic*.

Since valuation cannot be done in a vacuum, *clause 41(4)* requires the contractor to provide the quantity surveyor with any information which he requires to enable him to determine the value of any VI or determine any expense incurred through compliance with any other instruction. He must do this within 14 days of being asked for the information by the quantity surveyor: see clause 43(6).

Clause 42 Valuation of variation instructions

This clause provides two different methods whereby the value of a VI, as defined in clause 40(5), may be determined. These are:

(i) acceptance of a lump sum quotation meeting the requirements of clauses 42(2) and (3) and submitted by the contractor under clause 40(6); and

(ii) valuation by the quantity surveyor along conventional lines in accordance with the rules laid down in clauses 42(4) to (9): *clause 42(1)*.

Clauses 42(2) to (3) lay down the requirements where the PM has required a lump sum quotation for a variation under clause 40(6). Not every variation is suitable for valuation on a lump sum basis; this is a matter for the PM's decision but he will, no doubt, consult with the quantity surveyor before including such a requirement in a VI.

Clause 42(2) describes how the contractor's lump sum quotation is to be made up (see above) and requires him to submit supporting information to the quantity surveyor. If a lump sum quotation is required, the contractor must provide it within 21 days and in the manner specified in this sub-clause.

Under *clause 42(3)* the PM must notify the contractor within 21 days from receipt of his quotation whether or not it is accepted. Contractors should appreciate that if a lump sum quotation is submitted and accepted it is not subject to re-negotiation or re-valuation, e.g., because the actual disruption costs turn out to be greater than anticipated. Many lump sum quotations will, no doubt, contain a sizeable contingency element since the quotation as accepted 'shall be the full sum to which the contractor is entitled for complying with that VI'. The 21 day time limit may be extended by agreement under clause 1(4) and no doubt in practice any difficulties will be ironed out by negotiation before the quotation is submitted. Indeed, negotiation is envisaged by clause 42(3) which refers to the 'aggregate amount specified in [the] quotation, *or otherwise agreed between the Authority and the Contractor . . .*': see clause 48 (2)(b)

In practice, the quantity surveyor will carry out a detailed evaluation of the contractor's lump sum quotation and may himself undertake negotiations with the contractor: see clause 42(3). It is, however, for the project manager and not the quantity surveyor to make the decision as to whether to accept or reject the contractor's quotation.

Clause 42(4) applies both where no lump sum quotation is required and where the contractor's quotation is rejected or (in breach of contract) he

Instructions and payment 137

fails to submit a quotation. In those events, the quantity surveyor must value the variation in accordance with the rules in this sub-paragraph which are set out in order of application. There are important differences between these provisions and the corresponding provisions in clause 9(1) of Edition 2.

The valuation rules are:

- ***By measurement and valuation at the rates and prices in the bills for similar work: paragraph (a).***

The intention seems to be that the quantity surveyor must use this rule unless it is impossible for him to do so when he will move on to paragraph (b).

- ***By measurement and valuation at rates and prices deduced or extrapolated from the bill rates and prices: paragraph (b).***

The use of the word 'extrapolated' enables the quantity surveyor to deduce a rate which is outside the rates for similar work in the bills, as is apparent by reference to any standard dictionary. The PSA's Head of Claims has, at conferences, expressed the view that:

> 'You can only deduce a rate if you had a rate either side of the rate you were attempting to establish. In other words, if you had a rate for 100 and 150 mm slabs you can deduce a rate for a 125 mm slab. However, you could not deduce a rate for a 175 mm slab because it was outside the range. Extrapolation by definition means to extend beyond a known range'.

This is an accurate summary of the position. If it is not possible to value on this basis, the quantity surveyor moves on to paragraph (c).

- ***By measurement and valuation at fair rates and prices, having regard to current market prices: paragraph (c).***

The inclusion of the reference to current market prices is helpful in avoiding a common source of disputes.

If this is inapplicable there is a long-stop:

- ***By the value of the materials used and plant and labour employed in accordance with the basis of charge for daywork described in the contract.***

Effectively, because of the manner in which the clause is worded, the use of dayworks is restricted to those cases where it is impossible to measure the items of work involved, which again should reduce the scope for argument.

As noted already, clause 41(4) obliges the contractor to supply the quantity surveyor with the information he needs to value a VI. As the normal basis of valuation is the rates and prices in the bills, the quantity surveyor will need to break down the preliminary items. If there is insufficient information in the bills, the contractor must be prepared to provide further information, including a breakdown, especially where the quantity surveyor is being asked to evaluate the effects of disruption or prolongation.

Clause 42(5) enables the quantity surveyor to take account of the disruptive effect of the variation upon unvaried work. In doing so he will no doubt observe the guidance contained in the PSA Internal Technical Instruction TS 5/85. If the quantity surveyor forms the opinion that the VI has a disruptive effect on unvaried work, he must make the appropriate adjustment to the rates.

Clauses 42(6) and (7) impose time limits on both contractor and quantity surveyor. The contractor has 14 days in which to submit the information requested by the quantity surveyor under clause 41(4) and the quantity surveyor has 28 days from *receipt* of this information in which to notify his valuation to the contractor. Both these time periods can be extended by agreement under clause 1(4). If they are not, and the quantity surveyor fails to meet the time limit, the contractor would be entitled to finance charges under clause 47(1)(a).

Clause 42(8) is an interesting and unusual provision. If the contractor disagrees with the whole or part of the quantity surveyor's valuation he must notify the quantity surveyor within 14 days of the quantity surveyor's notification to him. He must give his reasons for disagreement and also his own valuation. If he fails to do anything, or merely disagrees without giving reasons and his own valuation, he is treated as having accepted the valuation 'and no further claim shall be made by him in respect of the VI'.

Clause 42(9) is a repetition of the proviso to clause 9(1) of Edition 2 and qualifies rules (a) and (b) in clause 42(4). It states that where an alteration or addition would otherwise fall to be valued under those rules 'but the QS is of the opinion that the relevant VI was issued at such a time or was of such a content as to make it *unreasonable* for the alteration or addition to be so valued', he shall use fair rates and prices.

It is the timing of the issue or content of the VI which triggers off this exception and the sub-clause confers upon the quantity surveyor a wide measure of discretion. The use of clause 42(9) is a matter for the professional judgment of the quantity surveyor.

Clause 42(10) deals with dayworks and (like clause 24 of Edition 2) requires the contractor to give reasonable prior notice to the quantity surveyor before commencing any work which the quantity surveyor has determined shall be valued as day work. The notice must be in writing (clause (1)(3)). Within one week of the week in which the work was done, the contractor must provide the quantity surveyor with supporting vouchers for verification. They are to be in the form required by the quantity surveyor and must give detailed accounts of labour, materials and plant for that pay week.

Clause 42(11) provides that if, as a result of a variation instruction, the contractor makes a *saving* in the cost of executing the works, the contract sum is to be decreased by the amount of any savings as determined by the quantity surveyor.

Clause 43 Valuation of other instructions

This clause is concerned with the valuation of all instructions other than variation instructions (VIs). It corresponds to clause 9 (2) of Edition 2. It is a two-edged sword and is almost identical to its predecessor. It deals primarily with the reimbursement of any 'expense' beyond that provided for in, or reasonably contemplated by, the contract and which the contractor *properly and directly* incurs as a result of an instruction of the PM. The restricted definition of 'expense' in clause 43(4) is important.

Clause 43(1) provides that where, as a result of an instruction other than the variation instruction the contractor either:

- properly and directly incurs any expense beyond that provided for in or reasonably contemplated by the contract; or
- makes any saving in the cost of executing the works;

the contract sum is to be increased or decreased, as appropriate, by the amount of that expense as determined by the quantity surveyor.

The use of the phrase 'properly and directly incurred' brings in the common law principle of remoteness of damage, and it must be emphasised that reimbursement of 'expense' under this provision is governed by the same principles as apply to damages at common law: *Wraight Ltd* v. *P.H. & T. (Holdings) Ltd* (1968). For a discussion of those principles readers are referred to *Building Contract Claims* by Vincent Powell-Smith and John Sims, 2nd edition (1989, BSP Professional Books, London). The expense must also be 'beyond that otherwise provided for in or reasonably contemplated by the Contract'. This imports an objective test; it is not the contemplation of the particular contractor.

Clause 43(2) is procedural and gives the contractor only 28 days from the date of complying with the instruction in which to submit to the quantity surveyor the information which he requires to enable him to determine the expense: see clause 41(4). This is a condition precedent to the operation of the clause and, under the terms of clause 41(3), already discussed, there will be no reimbursement of any expense where the instruction was necessary because of the default or neglect of the contractor or of any of his servants, agents or nominated or domestic subcontractors. An equally tight time-limit of 28 days is imposed on the quantity surveyor who must notify the contractor of his determination within 28 days of receipt of the relevant information. The failure of the quantity surveyor to comply with this time-scale will entitle the contractor to finance charges under clause 47(1), although of course these time periods may be waived by agreement: clause 1(4).

Instructions and payment

Clause 43(3) parallels clause 42(8) and requires the contractor to notify the quantity surveyor within 14 days of receipt of the quantity surveyor's determination if he disagrees with the amount determined. He must give reasons for his disagreement and his own estimate of the correct amount. If the contractor fails to respond or his response is defective, he is deemed to have accepted the quantity surveyor's determination and has no further claim.

Clause 43(4) defines 'expense' as meaning:

> 'money expended by the contractor [and it] shall not include any sum expended, or loss incurred, by him by way of interest or finance charges however described.'

In contrast to the situation under Edition 2, therefore, 'expense' is limited to money paid out, i.e., actually expended by the contractor.

Valuations and variations are illustrated in the flowchart Fig. 8.2.

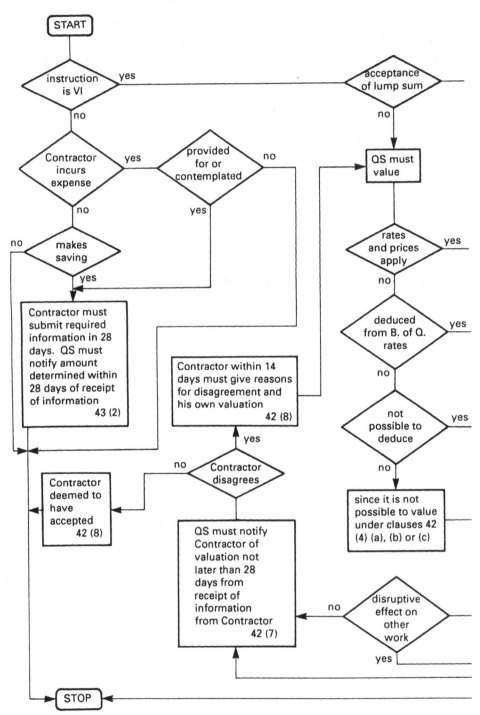

Instructions and payment

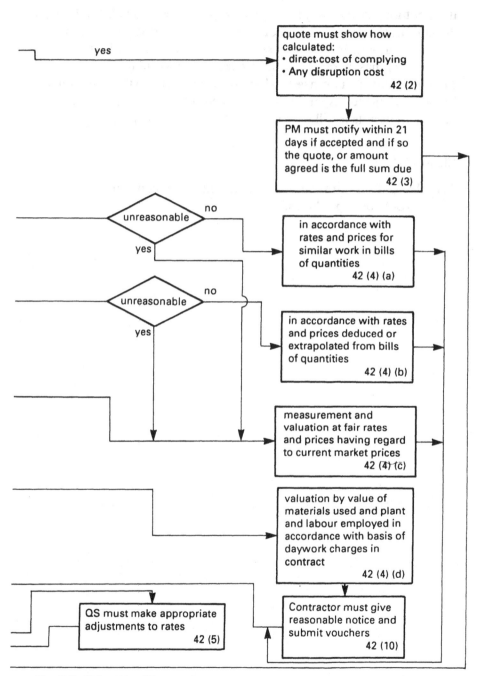

Fig. 8.2: Valuation of instructions

Clause 44 Labour tax

It is not thought necessary to discuss this provision in detail. It is largely the same as clause 11G of Edition 2 and is a fluctuation provision of the most limited type. It deals with statutory tax and contributions.

The PSA has a range of supplementary conditions dealing with variation of price for both general and specialist work on the basis of both formula (using indices) and traditional methods of calculation. If one of these is applicable it will be stated in the abstract of particulars.

Clause 45 VAT

This clause deals with value added tax, and any comment by a legal author would be superfluous.

Clause 46 Prolongation and disruption

Clause 46 deals with claims by the contractor for prolongation and disruption and is very limited when compared with claims clauses in other standard contracts, and indeed when compared with the equivalent clause in Edition 2. Variations, which are the main source of disruption and prolongation claims, are excluded, since the rules for valuing variations now allow for the disruptive effect by using the rates and prices in the bills of quantities, and by valuing the disruption consequent on the issue of a VI upon unvaried work as well: see p.136. Clause 46 is therefore restricted to 'breach of contract' type claims. Disruption can arise without prolongation and *vice versa*.

The subject of 'contractor's claims' is an emotive one and is often misunderstood. There are two main types of claim which can be made by the contractor:

- *Contractual claims, which are based on some express provision of the contract*, e.g., clause 46, the right to reimbursement under the contract being dependent on the contractor complying with the notice and related provisions of the clause.
- *Claims for breach of contract* and often called 'extra contractual claims'. These are claims for damages for breach of contract at common law and/or legally enforceable claims for breach of some other legal duty. Such claims may be based on breach of either express or implied terms.

The objective of this important clause is to reimburse the contractor for any expense which he *properly and directly incurs* in performing the contract and which he would not otherwise have incurred by reason of specified matters. Since the expense must have been 'properly and directly' incurred it means that any consequential loss is excluded. The distinction between direct and indirect or consequential loss was discussed by Mr Justice Atkinson in *Saint Line Ltd* v. *Richardsons, Westgarth & Co. Ltd.* (1940), in construing an exclusion of liability clause in a contract. It was there held that loss of expenses thrown away on wages and stores was recoverable as direct expense: see also *Croudace Construction Ltd* v. *Cawoods Concrete Products Ltd* (1978) where the Court of Appeal held that the cost of men and materials being kept on site was recoverable as damages and not excluded as 'consequential'.

The subject of prolongation and disruption claims is a large one and is treated in the various books devoted to the subject, including the volume which the author wrote with John Sims and entitled *Building Contract Claims* (*op. cit.*) to which readers are referred.

Under clause 46 the contractor is entitled to recover only 'expense' and this word is given a restrictive interpretation in sub-clause (6). It means money paid out by the contractor, and 'shall not include any sum expended, or loss incurred, by him by way of interest or finance charges however described'. Finance charges are not entirely excluded because there is a separate and limited provision in clause 47, but they are not permitted as a head of claim as they are under JCT 80, clause 26, for example. Furthermore, clause 46 is hedged about with restrictions and there are a number of hurdles which the contractor must overcome. There is also a very restricted list of events giving rise to claims.

Clause 46(1) and *clause 46(2)* must be read together to discover the events giving rise to claims. They are:

(a) The execution by the Authority of other work on the site at the same time as the works are being executed under clause 65: clause 46(1)(a).
(b) Any delay in the contractor being given possession of the whole or part of the site which unavoidably results in the regular progress of the works or part of them being materially disrupted or prolonged: clause 46(1)(b). The use of the word 'materially' clearly excludes trivial disruptions.
(c) Any delay in the respect of matters specified in paragraph (2) which unavoidably results in the regular progress of the works or any part of them being materially disrupted or prolonged. The specified matters are:

- Late design information which it is the PM's responsibility to provide: clause 46(2)(a). The actual wording is 'any drawings, schedules, levels or other design information to be provided by the PM'.
- Work or supply of goods which are to be ordered direct by the Authority or work or supply of goods to be undertaken by the Authority, except where this arises through the contractor's default: clause 46(2)(b). This ground is conditioned by *clause 46 (4)* since it is only applicable where the Authority has failed to supply an item or do something by a date agreed beforehand with the contractor or within any reasonable period specified in a notice given to the Authority or PM by the contractor.
- The Authority's or PM's instructions 'regarding the nomination or

appointment of or admission to the Site or issue of any pass to any person or any instruction or consent of the' employer under clause 63(2) (nomination): clause 46(2)(c).

These events are self-explanatory.

Clause 46(1) says that 'if the contractor properly and directly incurs any expense which he would not otherwise have incurred by reason of' one or more of the specified matters 'and which is beyond that provided for or reasonably contemplated by the Contract, the Contract Sum shall be increased by the amount of that expense as determined by the quantity surveyor.' It is not the function of the quantity surveyor to decide liability; his task is merely to calculate the sum of money (if any) due on the basis of the information submitted by the contractor in support of his claim or using information which he can reasonably be expected to have acquired in the course of his duties under the contract. This provision is hedged about with restrictions. The two most important are contained in *clause 46(3)* which makes it a condition precedent to any claim that:

- *Immediately* 'upon becoming aware that . . . regular progress . . . has been or is likely to be disrupted or prolonged' the contractor must have given written notice to the PM: clause 46(3)(a).
- This must specify the circumstances causing or expected to cause the disruption or prolongation *and* that the contractor is, or expects to be entitled to an increase in the contract sum under clause 46(1): clause 46(3)(b).
- As soon as reasonably practicable after incurring the expense, *and in any case within 56 days of incurring it,* the contractor must provide the quantity surveyor with *full details* of all expenses directly incurred and evidence that they directly result from one of the specified events.

This time limit has been criticised, but it must be remembered that that it may be extended by agreement under clause 1(4). Moreover, if for any reason the contractor does not make a claim under this provision, many of the events referred to, e.g., late information, also amount to a breach of contract at common law. GC/Works/1 does not exclude common law claims and consequently the contractor could, in the alternative, pursue his claim at common law: see *per* Mr Justice Vinelott in *London Borough of Merton* v. *Stanley Hugh Leach Ltd* (1985).

Clearly, however, clause 46(3)(b) requires the contractor to submit much supporting information and detail and the evidence provided must prove, on the balance of probabilities, that the actual expenses directly

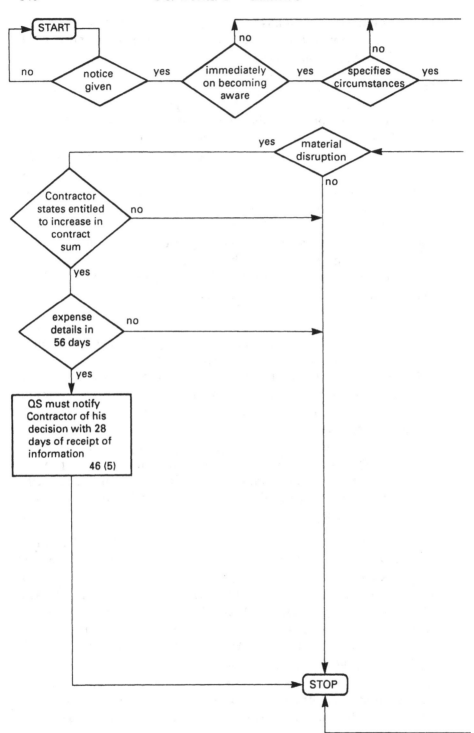

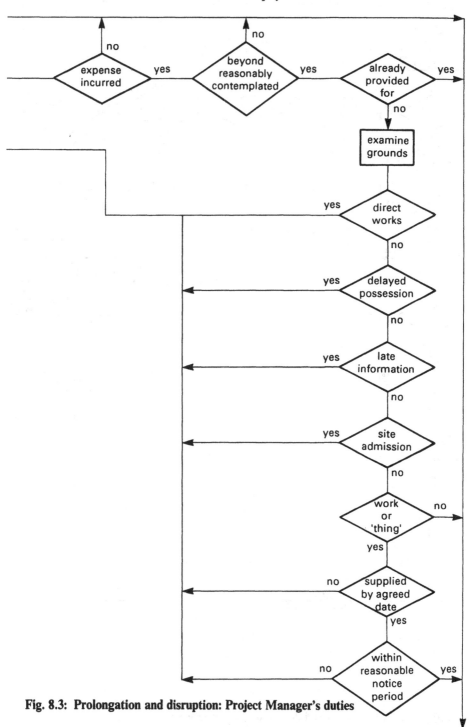

Fig. 8.3: Prolongation and disruption: Project Manager's duties

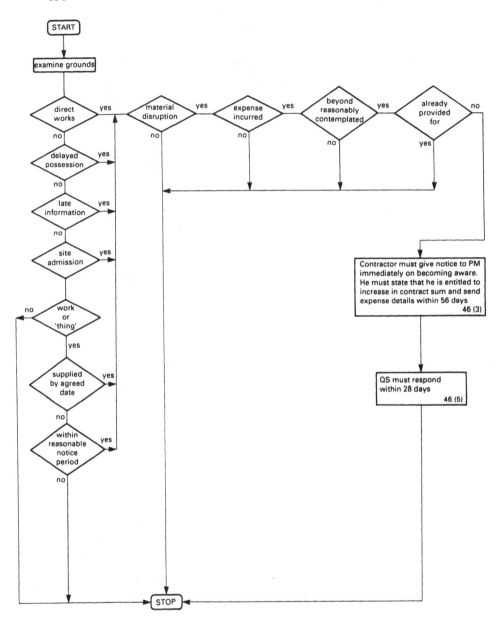

Fig. 8.4: Prolongation and disruption: Contractor's duties

Instructions and payment

result from the happening of one or more specified events. This would not, however, necessarily rule out 'global claims' or 'rolled up awards' in the *J. Crosby & Son Ltd* v. *Portland UDC* (1967) sense but, as pointed out in the *Leach* case, it is implicit in such a case that a global award:

> 'can only be made in a case where the ... expense attributable to each head of claim cannot in reality be separated and secondly, that [it] can only be made where apart from the practical impossibility the conditions which have to be satisfied before an award can be made have been satisfied in relation to each head of claim'.

The 'global approach' is equally applicable to extensions of time under clause 36.

In many cases the obligation imposed on the contractor by clause 46(3)(b) will be difficult to discharge, e.g., where the claim reflects expense incurred not only by the contractor but by his sub-contractors. In that case, agreement should be reached under clause 1(4) for the time to be extended.

Apart from claims for extra money under clause 46, provision is made in a number of conditions for payments to the contractor over and above the accepted tender price. These are set out in Table 8.2. Not all of them are 'claims' provisions but they bear on the subject as whole.

Clause 46(4) has already been mentioned. It is a condition precedent to

Table 8.2: Contract clauses providing for extra money

Clause	Content
3	Rectification of BQ errors
7	Unforeseeable ground conditions
11	Statutory undertakers' fees etc.
12	Royalties etc.
19	Loss or damage to property etc.
21	Defects in the works
31	Quality: Tests
33	Acceleration costs
41–43	Consequences of instructions
44	Labour tax charges
45	VAT
46	Prolongation and disruption
47	Financing charges
48	Monthly advances
50	Under-valuation
52	Cost savings
58	Determination costs
63	Failure of nominated sub-contractors

the contract sum being increased because of 'the execution of any work or the supply of any thing by the Authority or ordered from somebody other than the contractor' (clause 46(2)(b)) that the Authority has either failed to supply by a date agreed beforehand with the contractor, or within a reasonable period specified in a notice given by the contractor to the Authority or the PM for the supply of the item or taking the action.

Clause 46(5) imposes a duty on the quantity surveyor to respond to the contractor's claim within a short period of time. Unless the period is extended by agreement under clause 1(4) the quantity surveyor must notify the contractor of his decision on the claim within 28 days of receipt of all necessary information from the contractor, and if he fails to do so, the contractor will have an entitlement to finance charges under clause 47.

Clause 46(6) defines expense in the restrictive terms already noted above.

Sums determined by the quantity surveyor under clause 46 will be included in the next monthly advance: see clause 48(2)(d).

Flowchart Figs. 8.3 and 8.4 illustrate the procedure in relation to claims for prolongation and disruption.

Clause 47 Finance charges

It would be idle to pretend that the law relating to interest and finance charges is in a satisfactory state, but what is clear is that under Edition 2 of GC/Works/1 the word 'expense' included finance charges as one of its constituent parts. Finance charges in this context means the financial burden to the contractor of being stood out of his money. It covers interest actually paid out, e.g., on a bank overdraft, and also the loss of interest that might have been earned on money diverted from investment: *F.G. Minter Ltd* v. *Welsh Health Technical Services Organisation* (1980). There is also the possibility of interest being recoverable in limited circumstances as special damages, i.e., under the second limb of the rule in *Hadley* v. *Baxendale* (1854). The question is fully discussed in *Building Contract Claims*, 2nd edition (1989, BSP Professional Books, London) by Vincent Powell-Smith and John Sims, pp.145 to 157.

Subsequent to the *Minter* case, the PSA allowed finance charges as a permitted head of claim, but contractors and the Agency were in dispute over the correct interpretation of the law. Clause 47 seeks to remove all scope for differences of view, and must be read in conjunction with the definition of 'expense' in clauses 43(4) and 46(6), above, which specifically exclude 'interest or finance charges however described' from the definition.

In essence, clause 47 provides that when money is withheld from the contractor in specified circumstances which are the act or default of the Authority or its representatives an amount for finance charges will be added *automatically* to the relevant payments. The contractor does not need to provide any evidence that he has in fact incurred (or lost) interest.

Under the terms of clause 47(1), finance charges are payable to the contractor where the Authority, the PM or the quantity surveyor has failed to comply with any time limit specified in the contract and as a result money is withheld from the contractor; or where the quantity surveyor varies a decision of his which has been notified to the contractor.

Finance charges are payable as a percentage of the sums which would have been paid to the contractor at a rate of 1% over the rate charged during the relevant period by the Bank of England for lending money to the clearing banks and are added automatically to the money due.

Finance charges are not payable if they result from:

- any act, neglect or default of the contractor or any sub-contractor
- any failure by the contractor or any sub-contractor to supply the PM or quantity surveyor with any relevant information
- any disagreement about the final account.

The effect of clause 47(6) is to prevent the contractor from claiming interest as 'special damages' for breach of contract under the principles laid down in a series of cases culminating in *President of India* v. *Lips Maritime Corporation* (1988), CA and HL, and *Holbeach Plant Hire Co. Ltd* v. *Anglian Water Authority* (1988).

Clause 47(1) affords the contractor a right to interest or finance charges in specified circumstances. It says that the Authority shall pay the contractor finance charges 'only in the event that money is withheld from him under the contract' because the Authority, PM or QS has failed to comply with any time limit under the contract (clause 47(1)(a)) or the QS varies any decision of his which he has notified to the contractor (clause 47(1)(b)).

The rate of finance charges is prescribed in *clause 47(2)* as being 1% over the rate charged during the relevant period by the Bank of England for lending money to the clearing banks, and 'calculated as a percentage of the amounts which would have been paid to the Contractor if any of the' specified events had not occurred.

Clause 47(3) prescribes the periods during which finance charges are to be paid. This is from the date on which, 'but for a failure or variation [of a decision of the QS] . . . *money properly due under the contract became payable*, and ending with the date on which it is paid'. The italicised phrase gives scope for argument, but presumably it must be calculated by reference to the monthly periods for advances on account. (The first consultative draft of Edition 3 was more precise since it provided that, where paragraph (a) applied, the starting point was the date when the decision was required to be given or notified and, where paragraph (b) was applicable, from the date of giving or notifying the relevant decision to the date upon which the quantity surveyor gave notice to the contractor of the varied decision.)

A decision of the quantity surveyor may, of course, be varied subsequently by the adjudicator (clause 59) or in arbitration (clause 60), but no express provision is made to reflect this. The arbitrator, of course, enjoys a limited statutory power to award interest, but an adjudicator does not.

A change of decision by the quantity surveyor (clause 47(1)(b)) can in some circumstances result in a reduction of the amount due to the contractor. Accordingly, *clause 47(4)* provides that, when calculating finance charges, the quantity surveyor is to take into account any overpayment made to the contractor.

Clause 47(5) provides that the Authority is not liable to pay any finance charges which result from the contractor's own act, neglect or default or that of any of his sub-contractors, nominated or domestic, or by reason of any failure by him or his sub-contractors to supply the PM or QS with any relevant information or any disagreement about the final account. The contractor's failure, or alleged failure, to provide information, e.g., under clause 46(3)(b) may well prove to be a source of argument.

Finally, *clause 47(6)* precludes any possibility of the contractor seeking to recover finance charges or interest by way of special damages for breach on the part of the Authority and makes it clear that the contractor's only entitlement to finance charges is under the terms of clause 47 itself.

There is no definition of 'finance charges' in the contract and so the term must be interpreted in light of the principles laid down by the courts. In the first consultative draft there was a definition in the following limited terms:

'Finance charges means expenditure by the Contractor by way of interest or financial charges and do not include any financial loss of any description whether loss of profits, loss of use or opportunity, or however called.'

That definition was rightly criticised as being far narrower than that enunciated by the courts in such cases as *Minter* and *Rees & Kirby Ltd* v. *Swansea City Corporation* (1985) and as, indeed, being narrower than the PSA's own interpretation of finance charges under Edition 2. The acceptance of this criticism is presumably the reason for its deletion from Edition 3 as published.

The provisions of clause 47 appear to be very clear cut, and it seems that the only substantive criticism is that clause 47(3) sets a very low rate of interest which does not adequately compensate the contractor for being kept out of his money.

Clause 48 Advances on account

This clause, together with clause 49, sets out the scheme of payment of the contract sum to the contractor. It differs in many respects from the provisions found in most other standard forms of construction contracts which provide for instalment or interim payments.

The major innovation is that payment is made by reference to a stage payment chart or charts which will be included with the invitation to tender. The PSA has yet to finalise the form of the charts and a decision will have to be made between a chart using weeks on the vertical axis (Table 8.3) or a simplified version with percentage progress payments (Table 8.4).

Separate charts will be used for each nominated sub-contract, and the charts will not hold any retention which will be deducted on the face of the certificate.

Provided the contractor achieves the necessary rate of progress in accordance with the programme, he knows in advance the amount which he will receive each month, save for such matters as variations, reimbursement of prolongation and disruption expenses and finance charges. Since the contractor has the stage payment charts at the time of preparing his tender, he must allow in his tender price for any significant divergences between his resource costs and the stage payments.

Table 8.3: Stage payment chart showing effect of delay to project

Week no.	Original payment chart %	Delayed payment chart 1 %	Delayed payment chart 2 %
1	.4365		
2	.9484		
3	1.5337		
4	2.1901		
5	2.9156		
6	3.7081		
7	4.6655		
8	5.4856		
9	6.4664		
10	7.6057		
11	8.6014		
12	9.7515		
13	10.9538		
14	12.2061		
15	13.5065		
16	14.8527		
17	16.2427		

Instructions and payment

Week no.	Original payment chart %	Delayed payment chart 1 %	Delayed payment chart 2 %
18	17.6744		
19	19.1456		
20	20.6543		
21	22.1982		
22	23.7754		
23	25.3837	23.7754	
24	27.0210	25.3470	
25	28.6852	26.9475	
26	30.3741	28.5749	
27	32.0857	31.9020	
28	33.8179	33.5976	
29	35.5685	35.3120	
30	37.3355	37.0430	
31	39.1167	38.7887	
32	40.9100	40.5471	
33	42.7133	42.3160	
34	44.5245	44.0934	
35	46.3414	45.8774	
36	48.1621	47.6659	
37	49.9843	49.4568	
38	51.8060	51.2482	
39	53.6250	53.0379	
40	55.4392	54.8241	
41	57.2466	56.6045	
42	59.0450	58.3773	
43	60.8323	60.1404	60.1404
44	62.6064	61.8917	61.8114
45	64.3651	63.6292	63.4714
46	66.1065	65.3509	65.1186
47	67.8283	67.0547	66.7509
48	69.5284	68.7387	68.3666
49	71.2048	70.4008	69.9637
50	72.8553	72.0389	71.5403
51	74.4779	73.6510	73.0947
52	76.0703	75.2352	74.6248
53	77.6306	76.7893	76.1289
54	79.1565	78.3113	77.6050
55	80.6460	79.7993	79.0512
56	82.0970	81.2511	80.4657
57	83.5074	82.6647	81.8466
58	84.8750	84.0382	83.1919
59	86.1978	85.3694	84.4999
60	87.4735	86.6564	85.7685
61	88.7002	87.8971	86.9960
62	89.8757	89.0895	88.1805
63	90.9979	90.2316	89.3200
64	92.0647	91.3212	90.4127
65	93.0740	92.3565	91.4567

Table 8.3: Stage payment chart showing effect of delay to project (cont.)

Week no.	Original payment chart %	Delayed payment chart 1 %	Delayed payment chart 2 %
66	94.0236	93.3353	92.4502
67	94.9115	94.2557	93.3911
68	95.7356	95.1155	94.2777
69	96.4937	95.9129	95.1081
70	97.1837	96.6456	95.8803
71	97.8035	97.3118	96.5926
72	98.3511	97.9093	97.2429
73	98.8242	98.4362	97.8295
74	99.2208	98.8904	98.3504
75	99.5388	99.2699	98.8038
76	99.7761	99.5726	99.1877
77	99.9305	99.7966	99.5003
78	100.000	99.9397	99.7397
79		100.0000	99.9041
80			99.9915
81			100.0000

Notes: (a) The underlined week represents the week in which it has been agreed that valuations will take place. In this case the last Thursday of the month (The progress meetings being held on the previous Thursday.)

(b) Delay Chart 1 shows the effect of a one week delay recorded at the progress meeting held in week 25.

(c) Delay Chart 2 shows the effect of a further 2 weeks delay recorded at the progress meeting held in week 47.

Month	SUMMARY OF MONTHLY PAYMENT PERCENTAGES		
	Original payment chart	Delayed chart 1	Delayed chart 2
1	2.19%		
2	5.49%		
3	10.95%		
4	16.24%		
5	23.78%		
6	30.37%	30.23%	
7	37.34%	37.04%	
8	46.34%	45.88%	
9	53.63%	53.04%	
10	60.83%	60.14%	
11	69.53%	68.74%	68.37%
12	76.07%	75.24%	74.62%
13	82.10%	81.25%	80.47%
14	87.47%	86.66%	85.77%
15	93.07%	92.36%	91.46%
16	96.49%	95.91%	95.11%
17	98.82%	98.44%	97.83%
18	100.00%	99.94%	99.74%
19		100.00%	100.00%

Instructions and payment

Table 8.4: Simplified chart of percentage progress payments

% Complete	500-2000K	2000-3500K	3500-5500K	5500-7500K	7500-9000K	>9000K
1	.36	.20	.18	.10	.10	.11
2	.71	.39	.37	.20	.19	.23
3	1.16	.68	.64	.39	.39	.45
4	1.61	.97	.91	.59	.60	.68
5	2.14	1.34	1.28	.87	.90	1.01
6	2.68	1.71	1.64	1.16	1.20	1.34
7	3.29	2.17	2.08	1.54	1.60	1.77
8	3.91	2.63	2.53	1.92	2.01	2.20
9	4.60	3.16	3.06	2.39	2.50	2.73
10	5.30	3.70	3.58	2.85	3.00	3.26
11	6.07	4.31	4.19	3.40	3.58	3.88
12	6.83	4.92	4.79	3.95	4.16	4.50
13	7.67	5.61	5.47	4.58	4.83	5.21
14	8.51	6.29	6.14	5.21	5.50	5.92
15	9.42	7.05	6.89	5.91	6.25	6.71
16	10.32	7.80	7.63	6.61	7.00	7.50
17	11.29	8.62	8.44	7.38	7.82	8.37
18	12.26	9.44	9.26	8.16	8.65	9.24
19	13.28	10.32	10.13	9.00	9.54	10.18
20	14.31	11.20	11.00	9.84	10.44	11.13
21	15.39	12.14	11.94	10.74	11.40	12.14
22	16.47	13.08	12.87	11.64	12.36	13.15
23	17.60	14.07	13.86	12.61	13.39	14.23
24	18.73	15.06	14.64	13.57	14.42	15.30
25	19.91	16.11	15.88	14.59	15.50	16.44
26	21.08	17.15	16.92	15.61	16.59	17.57
27	22.30	18.24	18.01	16.68	17.72	18.76
28	23.52	19.33	19.10	17.75	18.86	19.95
29	24.78	20.47	20.23	18.87	20.05	21.19
30	26.04	21.60	21.36	19.99	21.24	22.43
31	27.33	22.77	22.53	21.16	22.47	23.72
32	28.62	23.95	23.70	22.32	23.71	25.00
33	29.94	25.16	24.91	23.53	24.98	26.33
34	31.26	26.37	26.12	24.73	26.26	27.66
35	32.61	27.61	27.36	25.97	27.57	29.02
36	33.96	28.86	28.60	27.22	28.89	30.38
37	35.33	30.13	29.87	28.49	30.23	31.78
38	36.70	31.40	31.15	29.77	31.58	33.17
39	38.09	32.70	32.44	31.07	32.95	34.59
40	39.47	34.00	33.74	32.37	34.32	36.01
41	40.88	35.32	35.06	33.70	35.72	37.45
42	42.28	36.64	36.38	35.03	37.12	38.90
43	43.69	37.98	37.72	36.38	38.53	40.36
44	45.11	39.32	39.06	37.74	39.95	41.82
45	46.52	40.67	40.42	39.10	41.38	43.29
46	47.94	42.03	41.77	40.47	42.81	44.77
47	49.36	43.39	43.14	41.86	44.26	46.25
48	50.79	44.76	44.50	43.24	45.70	47.73
49	52.21	46.13	45.88	44.63	47.15	49.21
50	53.63	47.50	47.25	46.03	48.60	50.70
51	55.04	48.88	48.63	47.42	50.05	52.18
52	56.45	50.25	50.01	48.82	51.50	53.67
53	57.86	51.63	51.39	50.23	52.95	55.15
54	59.26	53.01	52.77	51.63	54.40	56.62
55	60.65	54.38	54.14	53.03	55.85	58.09
56	62.04	55.75	55.52	54.43	57.29	59.56
57	63.41	57.12	56.89	55.82	58.72	61.01
58	64.78	58.49	58.26	57.22	60.15	62.47
59	66.14	59.84	59.62	58.61	61.57	63.90
60	67.49	61.20	60.98	59.99	62.99	65.33
61	68.81	62.54	62.32	61.37	64.39	66.74
62	70.13	63.88	63.67	62.74	65.79	68.15
63	71.43	65.21	65.00	64.10	67.16	69.53
64	72.72	66.53	66.33	65.46	68.53	70.91
65	73.99	67.84	67.64	66.79	69.88	72.25
66	75.25	69.14	68.94	68.13	71.23	73.60
67	76.47	70.42	70.23	69.44	72.54	74.91
68	77.69	71.70	71.51	70.76	73.85	76.21
69	78.87	72.95	72.77	72.04	75.13	77.48
70	80.05	74.20	74.02	73.33	76.41	78.74
71	81.19	75.42	75.25	74.58	77.64	79.96

Table 8.4: Simplified chart of percentage progress payments (cont.)

% Complete	500-2000K	2000-3500K	3500-5500K	5500-7500K	7500-9000K	>9000K
72	82.33	76.64	76.47	75.83	78.88	81.10
73	83.41	77.82	77.66	77.05	80.07	82.34
74	84.50	79.00	78.84	78.27	81.26	83.50
75	85.53	80.15	79.99	79.45	82.40	84.61
76	86.56	81.29	81.14	80.63	83.54	85.72
77	87.53	82.39	82.25	81.76	84.63	86.76
78	88.51	83.49	83.36	82.90	85.71	87.81
79	89.42	84.55	84.42	83.99	86.74	88.79
80	90.33	85.60	85.48	85.08	87.77	89.77
81	91.18	86.60	86.49	86.11	88.74	90.68
82	92.02	87.61	87.50	87.15	89.70	91.58
83	92.80	88.56	88.45	88.13	90.60	92.42
84	93.58	89.51	89.41	89.12	91.50	93.25
85	94.28	90.40	90.31	90.04	92.33	94.00
86	94.98	91.30	91.21	90.97	93.15	94.75
87	95.60	92.13	92.05	91.83	93.90	95.42
88	96.23	92.96	92.89	92.69	94.65	96.09
89	96.77	93.73	93.66	93.49	95.33	96.66
90	97.31	94.50	94.44	94.28	96.00	97.24
91	97.77	95.20	95.14	95.01	96.58	97.73
92	98.23	95.90	95.85	95.74	97.17	98.21
93	98.59	96.53	96.49	96.39	97.67	98.59
94	98.96	97.16	97.12	97.04	98.16	98.98
95	99.23	97.71	97.68	97.62	98.57	99.26
96	99.50	98.27	98.24	98.19	98.97	99.54
97	99.68	98.74	98.72	98.69	99.28	99.71
98	99.85	99.22	99.20	99.18	99.59	99.88
99	99.93	99.61	99.60	99.59	99.79	99.94
100	100.00	100.00	100.00	100.00	100.00	100.00

Clause 48(1) describes the contractor's basic entitlement. He is to be paid monthly advances on account during the execution of the works, subject to two provisos. First, that any relevant instructions have been or are being complied with. Second, that all the work to which the advance relates is to the satisfaction of the PM: see clauses 31 and 34.

At first sight this would suggest that if the work is not to the satisfaction of the PM in every respect, the contractor has no entitlement to the monthly advance and this is the strict interpretation of the provision. However, in practice what will happen is that if work has been executed which is not to the satisfaction of the PM it will be identified and an amount will be subtracted from the sum shown as otherwise due.

Under clause 50(1) the PM is to certify the sums to which the contractor is entitled under clause 48.

Clause 48(2) gives the breakdown of what each advance is to include, namely:

- 95% of the proportion of the sum specified in the stage payment chart for that month;
- 100% of any amount agreed under clause 42(1)(a) (variations — lump sum quotations) for the month;
- 100% of the agreed value or, failing agreement, 95% of the QS's

valuation under clauses 42(4) and clause 43 (traditional variation valuation and valuation of other instructions respectively);
- 100% of amounts determined by the QS for prolongation and disruption expenses under clause 46, and
- 100% of any finance charges payable under clause 47.

The balance of any sum which is less than 100% is accumulated by the employer as a reserve, i.e., as retention monies.

It should be noted that no retention is held where the value of a variation has been agreed (paragraphs (b) and (c)), and this must provide an incentive for contractors to agree, whether by way of lump sum quotation or under the conventional method of valuation.

Clause 48(3) provides for the PM to amend the stage payment chart to reflect actual progress and any changes resulting from any cost saving proposals made by the contractor under clause 52. The chart is to be adjusted where the PM has recorded in a statement after a progress meeting (see clause 35) that the works are in delay or ahead of programme. Fig. 8.5 shows a possible stage payment chart adjusted to reflect delay.

Clause 48(4) provides similarly for the charts to be amended where the Authority has accepted a cost saving proposal from the contractor under clause 52.

Clause 48(5) deals with credits for old materials. Any sums agreed to be credited by the contractor are deducted from the advance.

Clause 48(6) is no doubt intended to encourage contractors to pay their nominated sub-contractors and suppliers promptly. It reflects the provisions of Edition 2, clause 40(6). Its effect is that before payment of any advance, the PM can request the contractor to prove that any amount due to a nominated sub-contractor or supplier which has been included in a previous advance has in fact been paid, e.g., by way of a receipt.

Where the PM is not satisfied that payment has been made, then the Authority may withhold payment to the contractor of *the amount in question* until the PM is satisfied on this point. In the case of an unpaid nominated sub-contractor or supplier, the Authority retains a discretionary right to pay that person direct: see clause 63 (2).

Clause 47(7) provides for the Authority to accumulate the balance of any sum mentioned in clause 47(2) and which is less than 100%, i.e., under paragraph (a) 5% of the monthly proportion specified in the stage payment chart(s) and a like amount in respect of variations under paragraph (c) if their value has not been agreed. The 'reserve' is the retention fund and is not expressly stated to be a trust fund.

Clause 49 Final account

The provisions of clause 49 are a vast improvement on those of Edition 2 and provide a very clear procedure and a strict timetable for the preparation and agreement of the final account.

Clause 49(1) provides that when the works are completed to the PM's satisfaction, i.e., when he issues a certificate of completion under clause 39(1), the Authority must pay to the contractor *as soon as reasonably possible* the difference between the amount which the Authority estimates will be the final sum, less half the reserve, and the total amount of advances already paid. Thus, on completion the Authority revises its assessment of the final sum and half of the retention is released to the contractor.

Under *clause 49(2)* within six months of completion, the QS must forward his draft of the final account to the contractor showing how the final sum has been calculated. The contractor then has three months in which to notify his agreement or disagreement with the draft. If he disagrees, then he must state his reasons for so doing. If he fails to respond or disagrees without giving proper reasons, then *clause 49(3)* says that 'he shall be deemed to have agreed the draft final account as the Final Account'. If disagreement is properly notified then only negotiation or arbitration can result in a final sum.

Under *clause 49(4)* if the final sum is calculated and agreed before the end of the maintenance period, or in default of agreement has been determined by an arbitrator under clause 60, if the balance of that sum exceeds any reserve which the Authority is holding, the Authority must pay the excess to the contractor (paragraph (a)). In the converse case (paragraph (b)) the contractor must pay the excess back to the Authority.

Under the terms of *clause 49(5)* if, after the end of the longest specified period of maintenance, the PM has certified that the works are in a satisfactory state under clause 39(1), and the final account has been agreed (or deemed to be agreed under clause 49(3)) or determined by the arbitrator, the Authority must pay to the contractor any excess due or the contractor must repay any overpayment to the Authority.

Clause 50 Certifying payments

This straightforward provision deals with the issue of certificates by the PM and also says something about their natures and effect. It is based on clause 42 in Edition 2. Under the terms of *clause 50(1)* the PM must issue certificates in respect of sums to which the contractor is entitled under clauses 48 and 49. No form of certificate is prescribed, and all that is required, as Lord Justice Edmund Davies remarked in *Token Construction Co. Ltd* v. *Charlton Estates Ltd* (1973), is that:

'It must clearly appear that the document relied upon is the physical effect of a certifying process. One should, therefore, have some regard to the factors of "form", "substance" and "intent" of which by Mr Justice Devlin spoke in ... *Minster Trust Ltd* v. *Traps Tractors Ltd* (1954) ... The document should be "the expression in a definite form of the exercise of the ... opinion of the [PM] ... in relation to some matter provided for by the terms of the contract".'

Clause 50(2) deals with the nature and effect of certificates. Any interim certificate may be modified or corrected in any subsequent certificate or in the final certificate for payment. This would, in any case, be the position as regards interim certificates under the general law: see, e.g., *London Borough of Camden* v. *Thomas McInerney & Sons Ltd* (1986). By their nature interim certificates are merely approximate estimates and are not conclusive in favour of either party as regards either quantities or value of the work covered or as an expression of satisfaction with the quality of the work. The sub-clause then provides that 'no certificate of the PM shall of itself be conclusive evidence that any work or Things to which it relates are in accordance with the Contract' so that even the final certificate has no conclusive evidential effect and the problems which still arise under JCT 80, clause 30.9 are thereby avoided.

Clause 50(3) provides that any dispute as to the contractor's right to a certificate or as to sums to be certified is to be referred — at the contractor's request — to the Authority, whose decision is made final and conclusive.

A curious omission from this clause (and from the contract as a whole) is the fact that there is no provision for any period within which the Authority must honour certificates for payment. In the case of the PSA, this has not proved to be a problem in the past, but if the form were to be used in the private or local authority sector, an appropriate additional sub-clause would need to be inserted.

Clause 51 Recovery of sums

This short but important clause confers a contractual right of set-off on the Authority. This is a right which, in any event, the law would recognise in respect of the sums mentioned in the clause: see *Hanak* v. *Green* (1976). Such clauses are commonly found in standard building contracts and sub-contracts and their operation has given rise to an astonishing amount of litigation. These sorts of provision are an express contractual right and do not affect any other rights and remedies which the Authority may possess, e.g., in respect of defective work: *Acsim (Southern) Ltd* v. *Danish Contracting & Development Co. Ltd* (1989).

The position was stated clearly by Lord Diplock in *Gilbert Ash (Northern Ltd* v. *Modern Engineering (Bristol) Ltd* (1976) in the House of Lords:

> 'It is a principle of law [applicable to building contracts] ... stated authoritatively ... in *Mondel* v. *Steel* (1841) ... that when ... the person for whom the work has been done is sued ... by the contractor for the price it is competent for the defendant ... not to set-off ... but simply to defend himself by showing how much less the subject matter of the action was worth, by reason of the breach of contract ... This is a remedy ... available as of right ... independent of the doctrine of equitable set-off. ... in construing such a contract one starts off with the presumption that neither party intends to abandon any remedies for its breach arising by operation of law, and clear express words must be used in order to rebut this presumption ...'

There are no clear express words in GC/Works/1 which show that the Authority is abandoning any of its other rights and so clause 51 is merely an additional and most effective remedy.

Clause 51 is widely drawn. It allows the Authority to deduct sums 'recoverable from or payable by the Contractor' from 'any sum or sums then due or which at any time hereafter may become due to the Contractor under or in respect of the Contract or any other contract with any department or office of Her Majesty's Government'. The wording could scarcely be wider.

Clause 52 Cost savings

One of the draftsmen's stated aims when writing GC/Works/1. Edition 3, was to provide incentives for the contractor and this provision is a prime example of that philosophy. It is obviously designed to encourage contractors to explore and suggest the possibility of savings as the works progress. Clause 52 provides a contractual mechanism whereby the contractor can submit proposals for improved design, methods or components and, if his proposals are accepted, he shares the savings equally with the Authority.

Clause 52(1) provides that, at any time during the contract, the contractor may submit to the PM written proposals which in his opinion 'will reduce the cost of the Works or the cost of maintenanance or increase the efficiency of the completed Works'. The proposals must state that they are submitted for consideration under clause 52 and must include an estimate for consideration by the Authority.

If the Authority accepts the contractor's original or amended proposals, the contractor will be entitled to 50% of the saving.

Clause 52(2) requires the contractor to provide any further information about his proposals which either the PM or the Authority may require.

Clause 52(3) is an abundance of caution. It states that the contractor must continue with the expeditious carrying out of the works once he has submitted a cost saving proposal.

If the contractor's original or amended proposal is accepted by the Authority then, under *clause 52(4)*, the necessary amendments must be made to the date for completion and the programme (paragraph (a)) and the PM must award any necessary extension of time (paragraph (b)). The stage payment chart must also be adjusted under clause 48(4).

Chapter 9

Particular powers and remedies

Introduction

The penultimate section of the conditions contains eight clauses all of which deal with the special powers and remedies conferred on the Authority and the contractor. Its most interesting feature is the provision (clause 59) for *adjudication* on disputes which arise as the contract progresses, and the proper operation of this provision should do much to resolve disputes quickly and not allow them to fester so as to be harmful to co-operation and the project. The concept is derived from the BPF system (see *Manual of the BPF System*, pp.59–60: 1983, British Property Federation, London).

The other noteworthy feature is that the contract makes express provision for the power of determination to be exercisable only by the Authority. The Authority is also given a wholly discretionary right, which is not conditional on any act or default of the contractor, to determine the contract.

In contrast, the contractor is given no express right to terminate his own employment under the contract and must, therefore, rely on his common law rights. If, for example, the Authority were guilty of a serious breach of contract going to the very root of the bargain, the contractor would be entitled to accept that breach and treat the contract as at an end under the general law. This is a complex and difficult area of the law and reference may usefully be made to *Determination and Suspension of Construction Contracts* by Vincent Powell-Smith and John Sims (1985, Collins, London), especially chapter 1, which discusses the problems involved in depth.

A number of the contractor's powers are summarised earlier in Table 7.1 on page 121.

Clause 53 Non-compliance with instructions

Failure by the contractor to comply with instructions given by the PM under clause 40 is a breach of contract. This clause — which is modelled closely on clause 8 in Edition 2 — deals with the consequences of the contractor's failure to comply with the PM's instructions and provides the Authority with an immediate remedy. It is expressed to be 'without prejudice to the exercise of [the Authority's] powers to determine the contract': see clause 56(6)(a).

If the contractor fails to comply with an instruction of the PM, *clause 53* provides for the PM to issue a written notice (see clause 1(3)) to the contractor requiring him to comply with the instruction within the period specified in the notice. If the contractor does not then comply, the Authority is empowered to have the work done by others. The wording should be noted:

> 'the Authority may . . . . provide labour and/or Things (whether or not for incorporation), or enter into a contract for the execution of any work which may be necessary to give effect to that Instruction.'

The Authority is entitled to claim reimbursement from the defaulting contractor only to the extent laid down in the clause, i.e., it can recoup 'any costs and expenses incurred . . . *over and above those which would have been incurred had the contractor promptly complied with the instruction*'. These costs and expenses are recoverable by way of set-off under the provisions of clause 51.

Clause 54 Emergency work

This provision is based on clause 49 in Edition 2 and confers the necessary power on the PM to require the carrying out of emergency work, which is defined in sub-clause (3). It is not dissimilar to clause 62 of the ICE Conditions of Contract, 5th edition.

Clause 54(1) imposes a duty on the contractor to carry out any 'emergency work' required by the PM and instructed under clause 40(2)(k), while *clause 54(2)* provides an effective sanction. If the contractor fails to carry out the emergency work promptly, the Authority may make alternative arrangements for the work to be carried out, employing its own personnel or independent contractors. The Authority may recover from the contractor both the costs incurred (paragraph (a) and 'any loss suffered by the Authority because the contractor did not carry out the Work' (paragraph (b)).

Clause 54(3) defines emergency work as meaning 'any work which becomes necessary during the execution of the Works or during any maintenance period specified in the Abstract of Particulars':

- To prevent or alleviate the effects of any accident, failure, or other event in connection with the performance of the works. (The words 'other event' must be construed *ejusdem generis*.)
- To secure the works, the site or any adjoining property from damage. In the context it is thought that the 'damage' must be physical damage. (The wider definition of 'damage' in clause 19(6) is applicable only to that clause'.)
- To repair any damaged or dangerous part of the works. This is stated to be without prejudice to the provisions of clause 19 which requires the contractor, *inter alia,* (clause 19(2)) 'without delay and at his own cost to reinstate, replace or make good to the satisfaction of the Authority . . . . any loss or damage'.

Clause 55 Liquidated damages

This clause provides for the payment of liquidated and ascertained damages by the contractor if he fails to complete the works provided that a rate for liquidated damages is stipulated in the abstract of particulars. Where liquidated damages are agreed, they are exhaustive of the Authority's remedy for the breach of late completion: *Temloc Ltd v. Errill Properties Ltd* (1987). If there is a provision for liquidated damages and it fails for some reason, e.g., because it is in truth a 'penalty' or because the employer has prevented completion so making time at large the employer is left with his right to recover damages under the general law on the basis of his proven losses. The better view in those circumstances is that general damages which exceed the amount of the failed liquidated sum would not then be claimable and, although *Temloc Ltd v. Errill Properties Ltd* (1987) did not decide the point, it is a clear indication that the view expressed here is correct.

The object of a liquidated damages provision is to fix an amount which is a genuine pre-estimate of the employer's loss in the event of late completion or, in practice, a lesser sum, since the true figure would act as a disincentive to potential tenderers.

For a liquidated damages clause to operate successfully there must be a simple contractual machinery to produce a readily calculable mathematical result. This is provided by clause 55 and by clause 37 dealing with early possession.

The legal principles involved in liquidated damages are well settled, the most important point being that they are recoverable without the Authority having to prove loss. It is irrelevant if in the event there is no loss. This proposition has long been established law (*Clydebank Engineering Co. Ltd v. Castadena* (1905), HL) but is often challenged by contractors. A recent example is *BFI Group of Companies Ltd v. DCB Integration Systems Ltd* (1987), where Judge John Davies QC, Official Referee, held that it was quite irrelevant to consider whether there was any loss in fact. 'Much as I instinctively dislike provisions for liquidated damages,' he said, 'a provision of this sort is one which automatically comes into play once the event happens. There is no question here of it being a penalty, as the arbitrator had himself decided that it was not.'

Liquidated damages are often referred to erroneously as 'a penalty' but there is an important distinction between the two. A penalty is an extravagant sum of money and is irrecoverable. The dinstinction does not depend on how the parties describe the provision but on its *purpose*. In practice, the distinction may be difficult to draw on the facts, but if the provision is held to be a penalty it will be unenforceable.

To avoid being a penalty, the figure stipulated for must be a fixed and agreed sum which is a *genuine pre-estimate* of the likely loss to the Authority or a lesser sum. The genuineness of the estimate is to be judged at the date the contract is entered into and not at the time of the breach. A penalty, in contrast, is an extravagant sum of money inserted to coerce the contractor to performance: see *Dunlop Pneumatic Tyre Co. Ltd* v. *New Garage & Motor Co. Ltd* (1915) HL, for the classic guidelines laid down by Lord Dunedin. If a figure is held to be a penalty, then the Authority is entitled to sue for general (unliquidated) damages on the basis of his proven losses.

Clause 55(1) makes it clear that the condition is applicable only if a rate for liquidated damages is specified in the abstract of particulars. If none is specified, then the Authority's remedy for the breach of late completion would be by way of an action for unliquidated damages.

If the works or a section are or is not completed by the relevant date for completion then, under *clause 55(2)*, the contractor is *immediately* liable to pay the Authority liquidated damages at the specified rate 'for the period that the Works or any relevant Section remain or remains uncompleted'. The mere fact of late completion triggers off the operation of the provision. There is no requirement of a certificate of late completion or notice from the Authority to the contractor. The 'date for completion' is defined in clause 1(1) as 'the date or dates set out in, or ascertained in accordance with, the Abstract of Particulars or where extensions of time have been awarded, the date on which such extensions expire'. As to extensions of time, see clause 36.

Clause 55(3) confers an express right of deduction on the Authority. Any liquidated damages due to the Authority may be deducted from any advances to which the contractor would otherwise be entitled under clause 48. The Authority also has an express right of set-off under clause 51, which extends to setting-off against 'sums then due or which at any time thereafter may become due under [the contract] or any other contract with' the Government.

Clause 55(4) adverts to this right since it provides that if the amount of liquidated damages due to the Authority exceeds any advance payable to the contractor, the contractor must pay the difference to the Authority, and that amount is also recoverable under clause 51.

Clause 55(5) makes it plain that no waiver of liquidated damages can be implied. There must be an express waiver contained in a written (clause 1(3)) notice from the Authority to the contractor.

Clause 55 is a vastly improved and entirely redrafted version of clause

29(2) in Edition 2, and the elegant draftsmanship seems to overcome the problems often found in clauses of this type. It must be read in conjunction with the provisions as to extensions of time (clause 36) and early possession (clause 37).

Clause 56 Determination

Clauses 56 to 58 deal with determination of the contract and its consequences, and follow closely the pattern set by clauses 44, 45 and 46 of Edition 2, although there are significant differences between the two sets of provisions. It is quite clear that the express provisions for determination do not exclude the Authority's common law right to terminate the contract: *Architectural Installation Services Ltd* v. *James Gibbons Windows Ltd* (1989).

At common law one party may terminate the contract and recover damages if the other party fails to perform in a way which goes to the root of the contract. In other words, the breach of contract must be serious, i.e., repudiatory. The innocent party may also terminate the contract and recover damages if the other party states or shows by his conduct an intention not to perform the contract before the time for performance arrives. This is known as anticipatory breach of contract and the innocent party may terminate the contract immediately. He need not wait until performance falls due. For a discussion of the common law position see *Determination and Suspension of Construction Contracts* by Vincent Powell-Smith and John Sims (1985, Collins, London) pp.1 to 37; and Emden's *Building Contracts and Practice*, 8th edition, vol. 1, pp.250–257 (1980, Butterworths, London).

Clause 56 adds to the Authority's rights, and to some extent there is an overlap between the express powers of determination provided by the contract and the right to terminate at common law. In practice, the Authority will almost invariably proceed under clause 56 rather than at common law because clause 56 improves on the common law position and confers on the Authority a right to terminate the contract for breaches which would not so entitle it at common law.

Clause 56 is tightly drafted and gives the employer the right to terminate the contract not only for specific defaults but also without assigning any reason (see clause 56(7)), the latter power being necessary and justifiable in Government contracts but not in the hands of the private employer.

It is to be noted that the contract does not confer any express power on the contractor to terminate his own employment; any determination by the contractor must, therefore, be at common law.

Clause 56(1) provides that the Authority may determine the contract by notice to the contractor at any time. This power is stated to be 'without prejudice to any other power of determination', i.e., the Authority's powers at common law. No minimum or other length of notice is specified, but the notice must be in writing and the contract is determined upon receipt of the notice by the contractor: see clause 1(3).

GC/Works/1 is unusual in that it gives the Authority a *discretionary power* to terminate the contract without assigning any reason, in addition to the right to terminate for specified defaults. Hence *clause 56(2)* requires the Authority to specify in the notice which, of any, of the grounds specified in clause 56(6) apply.

The next three sub-clauses are procedural. The Authority is empowered to give directions to the contractor in relation to the performance or completion of any work and any other matters connected with the works, the site and any other contract or sub-contract: *clause 56(3)*. These directions must be given 'not later than three months from the date of the notice of determination (*not* its receipt) or the Date of Completion, whichever is the sooner': *clause 56(5)*. The contractor must comply promptly with such directions. He is to be paid for any work he performs in consequence on the same basis as an instruction, i.e., valuation under clause 43: *clause 56(5)*

Clause 56(6) specifies seven defaults by the contractor which entitle the Authority to determine the contract by notice under sub-clause (1). If notice is served because one of these defaults, post-determination matters are governed by clause 57.

The defaults are:

- contractor's failure to comply with an instruction within a reasonable period of its issue
- failure to execute work in an efficient, workmanlike or proper manner, or to proceed regularly and diligently with the works, or suspending their execution so that in the opinion of the PM, the work will not be completed on time
- insolvency and financial difficulties
- contractor's failure to comply with the contract provisions about admittance to the site (clause 26), where the employer decides that this is prejudicial to the interests of the State
- corrupt practices
- breach of other conditions mentioned in the invitation to tender.

Where the contract is determined for any other reason — which need not amount to a default — *clause 56(7)* provides that its consequences are to be dealt with under clause 58.

Under clause 30 goods or materials provided or used by the contractor to facilitate execution of the works but not for incorporation in them, e.g.,

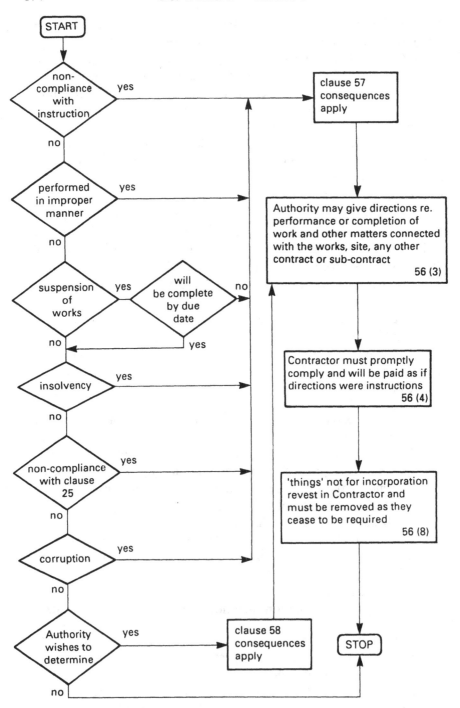

Fig. 9.1: Determination by employer

Particular powers and remedies

plant, tools, etc. are vested in the Authority for the duration of the contract. *Clause 56(8)* provides that, whether such things are damaged or not, they shall revest in the contractor and must be removed by him 'as and when they cease to be required in connection with any directions of the Authority given under': clause 56(3). From the date of determination, i.e., receipt of the determination notice by the contractor (see clause 56(1)) the Authority is under no liability to the contractor for any damage caused to them by the 'accepted risks': see clause 1(1) for a definition. In other words, these items are at the contractor's sole risk.

The flowchart Fig. 9.1 illustrates the determination procedure.

Clause 57 Consequences of determination for default

Clause 57 deals with the consequences of default determination.

Clause 57(1) states that the provisions listed apply. These may be briefly summarised:

- All sums of money then due or accruing due from the Authority to the contractor cease to be due.
- The Authority may enter on site and hire any person, employ other contractors, use any 'things on site' and purchase or do anything necessary to secure completion. (The contractor has no claim at all in respect of such action by the Authority.)
- Except in an insolvency situation, the contractor must assign without payment the benefit of any sub-contract or supply contract.
- The Authority may pay any sub-contractor or supplier any amount due and certified by the PM as included in a previous advance. Any amount so paid is immediately recoverable from the contractor.
- The cost of completion is to be certified by the PM and if there is (unusually) a credit due to the contractor, this would be paid to him after actual completion. More commonly there would be a deficit which would be payable by him: clause 57(3).

Clause 57(2) prescribes how the cost of completion is to be made up. It is to include:

- The value of all the work carried out to the satisfaction of the PM up to the date of determination.
- The value of any work etc. carried out pursuant to the Authority's directions under clause 56(3).
- The value of all 'things for incorporation' brought on to the site or in course of preparation or manufacture off site which the Authority elects to keep. This value is to be ascertained on the basis of fair and reasonable prices which are, presumably, the current market prices.

Clause 58 Consequences of other determination

The consequences of determination without reason or for a reason not listed in clause 56(6) are covered by clause 58 which (along with clause 56(7)) should be deleted if the form is used in the private sector, with the other necessary consequential amendments.

Clause 58(1) is purely declaratory and *clause 58(2)* sets out the financial consequences of a 'no fault' termination and provides the mechanism for the necessary calculation. In the calculation, in addition to the three items specified in clause 57(2), there is to be taken into account:

- Any reasonable sum expended by the contractor because of the determination of the contract in respect of:
  (a) the uncompleted part of any other sub-contract or contract, and
  (b) any unavoidable contract of employment entered into in connection with the contract

A similar balancing exercise to that under clause 56 is performed and the Authority either pays the contractor the excess or *vice versa*.

Clause 58(3) says that if, in the period from the date of determination (see clause 56(1)) to the 'date on which any directions under [clause 56(3)] are to have been complied with', an accepted risk causes damage to the whole or a part of the works or any items which the Authority has elected to keep, the amount payable to the contractor under the clause shall be ascertained as if no loss or damage had occurred, provided the contractor has properly performed his obligations with regard to protecting the works under clause 13.

Clause 58(4) gives the contractor who is the victim of a no fault determination a potential claim, but payment under this provision is entirely a matter for the Authority's discretion. It provides that, if the contractor is of the opinion that his unavoidable loss or expense, directly due to the determination, e.g., loss of profit, has not been fully reimbursed by any sums paid or agreed, he is to refer the matter to the Authority which 'shall make such allowance, if any, as in [its] opinion is reasonable'.

Unlike the corresponding provision in Edition 2, this is not stated to be a matter for the Authority's final and conclusive decision and so is reviewable in arbitration. It is a provision which empowers the Authority to make additional payments to the contractor in appropriate circumstances, since his financial entitlement under the clause as a whole is very limited indeed. It is understood that under earlier editions of GC/Works/1 the power to determine without reason was very rarely invoked, and in many ways its continued presence may be regarded as an historical hangover from the days of World War 2 and it is not believed that the discretionary power to determine 'without fault' is of frequent exercise.

Clause 59 Adjudication

One of the objectives of Edition 3 is to avoid disputes and to foster a team spirit. Clause 59 is an interesting attempt to prevent the 'running sore' syndrome which is all too prevalant when a dispute develops. It introduces *adjudication* as a means of disputes settlement during the contract period. Once again, it is based on the ACA/BPF model and, as printed in the conditions, would need amendment if the form were being used for private sector work. In that case, a provision for adjudication might well be based on clause 25 of the ACA Form of Building Agreement 1984, British Property Federation edition.

Clause 59(1) entitles the contractor to refer 'any dispute, difference or question arising out of, or relating to the Contract during the course of the Works' to adjudication to 'the person named in the Abstract of Particulars'. He is not, in fact, the adjudicator, but is to nominate the adjudicator: see sub-clause (3). The nominator in practice is a senior officer of the PSA.

It should be noted that the adjudicator has no jurisdiction in any dispute, difference or question arising where the contract makes the decision of the Authority or the PM 'final and conclusive'.

Moreover, a dispute cannot be referred to adjudication unless it has been outstanding for three months or more. The requirement for a cooling-off period implies that where there is a dispute the contractor and the PM should seek to resolve it amicably and it is submitted that the failure of either of them to do so is a matter which the adjudicator should take into account under sub-clause (6). Only the contractor can ask for adjudication and he can do so on 'any matter which has been outstanding for three months or more', thus implying that there is an obligation on the parties to seek to resolve disputes amicably.

Clause 59(2) deals with the content of the contractor's notice. It must specify the matter in dispute and set out the principal facts and arguments relating to it. All relevant documents which the contractor has in his possession must be attached to the notice. When sending the notice to the person named, the contractor must send a copy of it and its annexures to both the PM and the quantity surveyor.

Clause 59(3) provides for the appointment of the adjudicator. When he receives the contractor's notice, the named person must:

(a) nominate an officer of the Authority or person acting for the Authority to act as an independent adjudicator. The person nominated shall not have 'been associated with the letting or management of the Contract'; and,

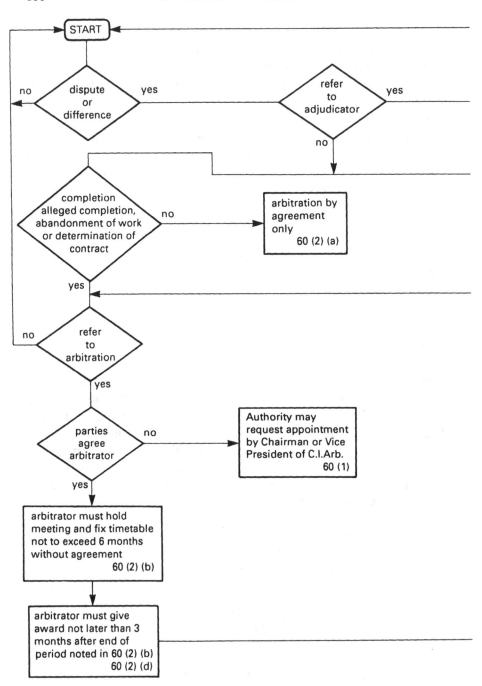

Particular powers and remedies

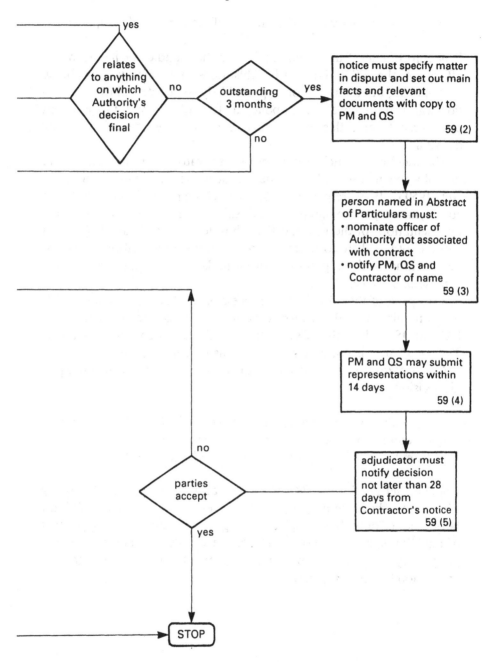

Fig. 9.2: Adjudication and arbitration

(b) notify the contractor, PM and QS of the name of the adjudicator.

It is unfortunate, if understandable, that the adjudicator has to be an officer of the PSA. Obviously, this does not prevent him being independent, but a number of disputes likely to be referred to adjudication will turn on basic general policy matters of PSA and Government procurement, and this must inevitably influence the adjudicator's decision.

The adjudicator is not in any sense an arbitrator in law; his role is more that of a conciliator, and consequently the provisions of the Arbitration Acts 1950 to 1979 are inapplicable. In reaching his decision he must, of course, have regard to the contractor's submissions (clause 59(2)) and any representations of the PM and QS submitted under *clause 59(4)* which entitles both of them to submit representations (presumably these must be in writing) to the adjudicator within 14 days of receipt by them of the contractor's notice.

The speed of the adjudication process is emphasised by *clause 59(5)* which requires the adjudicator to notify his decision to the contractor, PM and QS not later than 28 days from receipt of the contractor's notice. The adjudicator is not obliged to give reasons for his decision.

Clause 59(6) is important since it directs the adjudicator in coming to his decision to:

'have regard to how far the parties have complied with any procedures in the contract relevant to the matter in dispute *and to what extent each of them has acted promptly, reasonably, and in good faith.*'

Under the terms of *clause 59(1)* the adjudicator's decision is made binding while the works are in progress, i.e. it is given temporary finality. It 'may be questioned only after completion, alleged completion or abandonment of the Works or the determination of the Contract'. Thereafter, either party may seek arbitration under clause 60 unless they agree to arbitration before completion.

Clause 60 Arbitration

The arbitration agreement (clause 60) contains some unusual features, not least a strict timetable for the arbitrator to follow. This envisages his holding a preliminary meeting 'forthwith on his acceptance of office' and a procedural run-in of not more than six months from that meeting. The arbitrator is to give his award within a further three months and so in theory a maximum period of nine months should elapse before the arbitration has run its course. It bears little remembrance to the corresponding provision (clause 61) in Edition 2.

Clause 60(1) provides that 'in addition to adjudication' under clause 59 disputes may be settled by reference to arbitration. It sets out the scope of the arbitration agreement. Arbitration is available on all 'disputes, differences or questions between the Authority and the Contractor arising out of or relating to the Contract'. The exception is that arbitration is not available where a decision is expressed to be 'final and conclusive'.

The following decisions are stated to be final and conclusive and are thus excluded from review by the arbitrator:

(a) Clause 6 (1) — Contractor's employees. By clause 6(1) any decision of the PM about the replacement of the contractor's employees is stated to be final and conclusive.

(b) Clause 24(3) — The Authority's decision on matters of corruption etc. is, by clause 24(3), made final and conclusive.

(c) Clause 26(3) — This sub-clause makes the decision of the PM as to whether any person is to be admitted to the site etc. final and conclusive.

(d) Clause 31(6) — Independent expert's reports on tests are made final and conclusive.

(e) Clause 37(7) — The decision of the SO on matters of early possession is final and conclusive.

(f) Clause 39(2) — The Authority's decision on any dispute as to the contractor's right to a completion certificate or one that the works are in a satisfactory state is made final and conclusive by this sub-clause.

(g) Clause 40(3) — PM's instructions. The decision of the PM as to

whether instructions are necessary or expedient is made final and conclusive.

(h) Clause 44(5) — The decision of the Authority about the amount of increase or decrease in the contract sum on account of labour tax fluctuations is stated to be final and conclusive.

(i) Clause 50(3) — The decision of the Authority about the contractor's right to a payment certificate or as to the sum to be certified is made final and conclusive, but this does not apply to a dispute as to the amount of the balance of the final sum due to the contractor.

(j) Clause 59 — Determination. The Authority's decisions on all matters arising under clause 59 are final and conclusive: clause 59(9)

(k) Clause 63 — The decision of the Authority as to the reasonableness or otherwise of the contractor's objection to a proposed nominated sub-contractor is made final and conclusive by clause 63(6).

(l) Clause 18(3) — Measurements taken by quantity surveyor where the contractor's representative fails to attend are 'final and conclusive' for the purposes of the final account.

Apart from these twelve exclusions — which have been reduced in number when compared with those in Edition 2 — the clause is very wide and covers all matters arising out of the contract, although interestingly the arbitrator is not given an *express* power to 'open up, review and revise' the decisions, opinions etc. of the Authority, the PM and the quantity surveyor; such power is implied.

The machinery for bringing the clause into operation is straightforward. All that is required is a written notice (see clause 1(3)) of the dispute or difference from either party to the other, seeking concurrence in the appointment of an arbitrator. Either party may seek arbitration and if notice of arbitration is given both sides must make reasonable efforts to agree on a single arbitrator. The Chartered Institute of Arbitrators, 75 Cannon Street, London EC4N 5BH will provide the names of suitably qualified arbitrators on request.

In default of agreement within a reasonable period — which, it is suggested, is no more than a month — then the arbitrator is to be appointed *at the request of the Authority only* by the Chairman or Vice President of the Chartered Institute of Arbitrators.

Clause 60(2) is prefaced by the words 'unless the parties otherwise agree' and (paragraph (a)) places a limitation on the commencement of arbitration. Notice to concur in the appointment of an arbitrator cannot be given until after the completion, alleged completion or abandonment of the works or the determination of the contract. This is a standard provision in clauses of this type and because of the provision for adjudication (clause 59) is unlikely to cause difficulty.

The remaining paragraphs of clause 60(2) are procedural. Under paragraph (b) the arbitrator is to hold a preliminary meeting with the parties 'forthwith on his acceptance of office'. At that meeting, he is directed to set a timetable for the delivery of the points of claim and defence or counter-claim, for discovery and inspection of documents, for any necessary inspection of the works, and for the hearing of oral evidence if necessary. These are the usual matters dealt with at a preliminary meeting, and what is unusual is the statement that the 'timetable shall not without the consent of the parties exceed six months' from the date of the preliminary meeting.

Obviously, it is essential that, before he is appointed, the arbitrator agrees to accept as part of· his conditions of engagement that the appropriate procedures and time limits will apply unless the parties agree otherwise.

Paragraph (c) requires the parties to ensure that any evidence, documentary or otherwise, is submitted to the arbitrator in accordance with the agreed timetable.

Finally, under paragraph (d) the arbitrator is enjoined to give his award not later than nine months from the date of the preliminary meeting. If this timetable can be adhered to, speedy resolution of disputes will result, but in practice it is to be anticipated that the proceedings will be much longer.

Clause 60(3) says that where the contract is subject to English law, the reference to arbitration is one to which the Arbitration Acts 1950 to 1979 apply. The arbitration will, therefore, be governed by those Acts and the proper law of the contract (procedural and otherwise) will be English law. The Arbitration Acts 1950 to 1979 provide many important procedural and other advantages: see, for example, *Handbook of Arbitration Practice*, edited by R. Bernstein (1987, Sweet & Maxwell, London), Part 2.

Clause 60(4) covers the situation where the contract is governed by Scots law, when different legal considerations apply.

The flowchart in Fig. 9.2 illustrates the adjudication and arbitration procedure.

Chapter 10

Assignment, sub-letting, sub-contracting, suppliers and others

Introduction

Unlike other standard forms of contract, GC/Works/1 has always dealt with sub-contractors comparatively briefly, since its draftsmen believe that their progress and work should be the main contractor's responsibility. There has, however, been some slight amelioration in respect of the contractor's liability for nominated sub-contractors.

The five clauses in this section deal not only with assignment, sub-letting and nomination, but also with the carrying out of work on site by others at the same time as the contract works are being executed. Similar provisions in other forms of contract have given rise to a body of case law.

Clause 61 Assignment

Under the general law of contract, contractual duties and liabilities cannot be assigned. The only practical effect of this provision is, therefore, to prevent the contractor from assigning or transferring his contractual rights to a third person. Assignment is prohibited unless the Authority gives its written consent. However, the law draws a distinction between assignment of liabilities under a contract and vicarious performance of contractual liabilities, i.e., someone else performing the obligation whilst the contracting party remains fully liable under the contract — in other words, sub-letting. This is dealt with separately by clause 62.

Clause 61 clearly prevents the contractor from raising finance by assigning his rights to payment under the contract unless the Authority consents in writing: see on this point generally *Helstan Securities Ltd* v. *Hertfordshire County Council* (1978). The last sentence of the clause says that 'no sum of money shall be payable to any person other than the contractor without [the contractor's] written consent'. This would not, it is submitted, be effective against a garnishee order, i.e., an order of the High Court requiring a third party who owes money in the ordinary course of business to a judgment debtor to pay the amount direct to a judgment creditor: see *Dunlop & Ranken Ltd* v. *Hendall Steel Structures Ltd* (1957).

Clause 62 Sub-letting

It is often said that the bulk of actual construction work in the United Kingdom is in fact carried out by sub-contractors, whether nominated or domestic, and the opening sentence of *Default by Sub-contractors and Suppliers* by John Parris (1985, Collins, London) states baldly that: 'Sub-contractors and building material suppliers are the Achilles' heel of the British construction industry.' They have certainly caused more than their fair share of legal problems.

It is extremely important to appreciate that the doctrine of privity of contract means that only the parties to a contract can acquire rights or liabilities under it. Thus, in a sub-contract situation the only contractual relationship is between the main contractor and the sub-contractor, whether that sub-contractor be a nominated or domestic one. The result of this is that the main contractor carries the legal responsibility for the sub-contractor's work so far as the Authority is concerned. It is to the main contractor that the Authority will look if something goes wrong. The Authority cannot sue the sub-contractor direct for bad workmanship or other defaults — unless there is a collateral warranty or other direct contract existing between them — and conversely the sub-contractor cannot sue the Authority direct.

Each party has contractual rights only against the party above or below him in the contractual chain. The Authority is, therefore, able to claim against the main contractor for defective work done by a sub-contractor, or, indeed, for any default amounting to a breach of sub-contract on the part of that sub-contractor, because the sub-contractor's failure is also a breach of the terms of the main contract. In turn, the main contractor may sue the sub-contractor or join him in the proceedings as a third party and in this way claims are passed up and down the contractual chain.

In the present state of the law it is extremely doubtful whether the Authority will have any direct claim in tort against the sub-contractor, e.g., for damage caused by negligently bad work: *D & F Estates Ltd* v. *Church Commissioners for England* (1988); *Simaan* v. *Pilkington Glass Ltd* (1988); and *Greater Nottingham Co-operative Society* v. *Cementation Piling & Foundations Ltd* (1988). The case of *Junior Books Ltd* v. *Veitchi Co. Ltd* (1983) is effectively dead, although it has yet to be decently interred.

The express provisions in GC/Works/1 dealing with sub-letting are, therefore, of great importance, and GC/Works/1 proceeds on the fundamental assumption (which is also the position at common law) that the contractor is fully responsible for his sub-contractors. Clause 62 is based on clause 30 of Edition 2 and there is no change in principle. It is concerned with *vicarious performance* of the contractor's duties. Vicarious performance is normally permitted except where the nature of the contract calls for personal performance which will not, it is submitted, normally be the case in contracts for building and civil engineering works. However, vicarious performance is only effective to discharge the contractor's duties if it is perfect.

Clause 62(1) says that the contractor must not sub-let any part of the contract without the previous consent of the PM. However, the opening sentence of the sub-clause recognises that the main proposed domestic sub-contractors may have been approved by the Authority at tender stage and that the contract may specify or nominate the sub-letting of work. Any further sub-letting requires the previous consent of the PM. If consent is sought the contractor must provide him with 'such details of any sub-contractor he wishes to engage as the PM may require'. The PM's consent is not subject to any qualification that it must not be unreasonably withheld. Consent will depend upon the PM accepting that the proposed sub-contractor is competent to carry out the work it is proposed to sub-let, but there is no contractual requirement that the sub-contractor should be on the PSA's approved list.

Clause 62(2) does not specify the detailed terms of any domestic sub-contract. The way in which the main and sub-contract fit together has proved a major practical problem in the past and provisions of the main contract cannot be read into the sub-contract unless they have been expressly incorporated: *Smith and Montgomery* v. *Johnson Brothers Co. Ltd* (1954).

There is at present no sub-contract form which is tailor-made for Edition 3, and the existing GW/SC published by the Building Employers Confederation for use with Edition 2 would need some amendment, pending the issue of a revised edition.

Clause 62(2) imposes a general obligation on the contractor to step-down to the sub-contractor rights and obligations similarly to those enjoyed by or imposed on the main contractor under his contract, and to give to the sub-contractor similar rights and remedies against him as he enjoys against the Authority.

However, every sub-contract must contain certain minimum terms:

- Provision to determine the sub-contract if the Authority determines the main contract under clause 56.
- A vesting clause in respect of the sub-contractor's materials etc. 'to the effect that from the commencement to the completion of the sub-

contract all things belonging to [the sub-contractor] which are brought on site in connection with the sub-contract shall vest in the Contractor subject to any right of the contractor to reject the same': see clause 30.
- Terms enabling the main contractor to fulfil his obligations to the Authority under clauses 2(4) (contract documents), 22 (government premises), 26 (site admittance), 27 (passes), 28 (photographs), 30(2) (vesting), 31(2),(3) (quality).
- Terms imposing on the sub-contractor liabilities similar to those imposed on the contractor by clause 6 (contractor's employees), 24 (corruption), 29 (secrecy), and 61 (assignment)
- A term aginst further sub-letting without the contractor's consent.
- Where the sub-contract inmvolves the execution of work on the site a fluctuations clause in terms of clause 44(1) and (2).

Clause 62(3) is stated to be 'without prejudice to the obligations of the contractor under any of the provisions of the Contract' and obliges the contractor, if requested to do so by the Authority, 'to take any necessary action to ensure that [a sub-contractor] complies with and performs all obligations imposed upon him'.

Clause 62(4) makes the contractor's position crystal-clear:

'Where for any reason a sub-contract is determined ... because of the default or failure of the sub-contractor, the Contractor shall, subject to Condition 63(9) (*Nomination*), at his own expense secure completion of the sub-contract works.'

This echoes the common law position as stated by Lord Justice Collins in the old case of *Mitchell* v. *Guildford Union (1904)*:

'The contractor ... has accepted the primary obligation of completing the work within the given time ... the contractor has accepted ... as between himself and the owners, the primary obligation in respect of the sub-contractor's work just as much as with regard to the other'.

Clause 63 Nomination

Clause 63 is modelled on the similar provisions in Edition 2 and, in the writer's view, it is a manifestly better provision than the convoluted clause 35 of JCT 80. In essence, once a sub-contractor is nominated and accepted by the main contractor, he is the main contractor's responsibility. There is no right to extension of time if a nominated sub-contractor causes delay or fails.

Edition 3 maintains the Edition 2 option of nomination using prime cost items, but nominated sub-contractors could also come about through naming a single specialist in the specification, for example, or giving a list of alternatives.

Clause 63(1) gives a definition of both nominated sub-contractors and nominated suppliers. Those terms mean:

'a person with whom the Contractor is required to enter into a contract for the execution of work or the supply of Things designated as "Prime Cost" or "PC" items. *This requirement may be specified in the Contract documents or in any direction or instruction under the contract.*'

A nominated sub-contractor or supplier can, therefore, be nominated in the contract documents (in which case the contractor has no right to object to the nominee: see clause 63(6)) or else be named in any direction of the Authority or instruction issued by the PM against a PC item.

Clause 63(2) confers a power of nomination on the Authority and the PM. All PC items are reserved for carrying-out by nominated sub-contractors or suppliers ('persons to be nominated or appointed in such ways as may be directed ... or instructed ...'). The contractor must not order PC work or other PC items before an 'authorised sub-contract' is made. The final sentence of the sub-clause reserves to the Authority the right to order and pay for all or any part of a PC item direct. Where the Authority does so contract direct, then the Authority may deduct 'these payments' from the contract sum 'less an amount in respect of contractor's profit at the rate included in the Bills of Quantities adjusted pro-rata on the amount paid direct by the Authority' to the sub-contractor or supplier: see clause 46(2)(c) and (4) as to recovery of the cost of prolongation and disruption by delay in the exercise of this power and clause 35(2)(b) as to extension of time

Under *clause 63(3)* the sum to be paid by the Authority in respect of any PC work or item is the sum properly due to the sub-contractor or supplier, after adjustment in respect of overpayment, etc. The sum payable includes proper charges for packing, carriage and delivery to site. Any increases or decreases in the PC sum included in the contract are to be added to or deducted from the contract sum. The contractor must provide the quantity surveyor with the necessary quotations, receipted invoices and bills necessary to show the sums actually paid.

Clause 63(4) states that the contractor is also entitled to payment for fixing items supplied by a nominated supplier in accordance with the rates included in the bills and to profit. The contractor's profit at the rate included in the bills is adjusted pro rata on the prime cost, excluding any alterations in the prime cost due to the effect of any fluctuations clause in the sub-contract. This is the meaning of the reference to 'conditions incorporated . . . . pursuant to Condition 62(2)'. The payment for fixing is to cover unloading, getting-in, unpacking, return of empties and other incidental expenses.

Where any work or 'things for incorporation' are subject to a nominated sub-contract but are not included in the bills, *clause 63(5)* says that, if required to do so by the Authority, the contractor must supply a full and detailed schedule of rates 'which was properly and reasonably used for calculating the Contract sum or sub-contract sum'. That schedule is then to be used in place of the bills for the purposes of measurement and valuation of the work or item in question.

Clause 63(6) confers on the contractor a right to make *reasonable objection* to a proposed nominated sub-contractor or supplier *except where he is nominated in the contract documents provided by the Authority as the basis of the contractor's tender*:

> 'The Authority shall not require the Contractor to enter into a sub-contract with any sub-contractor against whom the contractor has made reasonable objection . . . .'

The contractor is bound to provide the PM with such information as he reasonably requires in relation to his objection, and the Authority's decision as to whether the objection is reasonable is made 'final and conclusive'. If the contractor wishes to object, it is suggested that he must do so at the time the nomination instruction is issued.

'Reasonable objection' is not defined, but it is submitted that it includes both questions of technical competence and financial reliability; incompatibility of the sub-contractor's programme with the programme would clearly be a reasonable objection (see *Percy Bilton Ltd* v. *Greater London Council* (1980)).

The contract is silent as to what is to happen if the contractor makes a

Assignment, sub-letting, sub-contracting, suppliers and others 193

reasonable objection which is sustained; presumably there must be a renomination and this might, or might not, dependent on the circumstances, give rise to a financial claim under clause 46 and to an extension of time under clause 36.

Under *clause 63(7)* the contractor may only terminate a nominated sub-contract with the agreement of the Authority. Where the Authority so agrees, the Authority must as soon as is reasonably practicable either nominate a replacement *or* direct the main contractor to complete the work either by his own resources or by a sub-contractor of his choice, approved by the Authority. If, in these circumstances, the contractor proposes the use of a sub-contractor — as would generally be the case — the PM's consent to the sub-letting would be required under clause 62(1).

Delay consequent on the termination is not a ground for extension of time and, subject to one exception (see below), the financial consequences of the failure are also the contractor's responsibility.

Although the sub-clause provides for renomination, there is no duty on the Authority to renominate as is the case, for example, under JCT 80, clause 35. This is an unusual provision to find in a contract of this sort, for, as the House of Lords made clear in the well-known case of *Northwest Metropolitan Hospital Board* v. *T.A. Bickerton Ltd* (1970), under the normal scheme of nomination, the main contractor cannot be required to do the work of a nominated sub-contractor in the event of failure by the nominated sub-contractor. Under clause 63(5) of GC/Works/1 the contractor can be required to do the work and generally at the original price. This is made plain by *clause 63(8)* which reads:

'Subject to paragraph (9) if a nominated subcontract is determined or assigned the Authority shall not be required to pay the Contractor any greater sums than would have been payable if determination or assignment had not occurred.'

Clause 63(9) introduces an important exception to the principle that the Authority is only required to pay the sum which would have been payable had the nominated sub-contractor not failed. If the reason for the failure is *insolvency* then the Authority is to reimburse the contractor

'an amount equal to the difference between —
  (a) any cost he has incurred in securing the completion of the sub-contract works which exceeds the cost to him of completing the Works under the original sub-contract, and

(b) the amount which, having used his best endeavours, he has or should have recovered from the original sub-contractor.'

In simple terms, this means that the contractor will recover from the Authority any shortfall which he has not recovered from the original sub-contractor in liquidation or receivership. This contrasts favourably with the position under Edition 2.

Clause 64 Provisional sums

This clause is identical with clause 39 in Edition 2. It deals with the expenditure of 'provisional *lump* sums' and 'provisional items inserted in the Bills', and at first sight the terminology is confusing. In fact, it is suggested that nothing turns on the use of the phrase 'provisional lump sum' which, it is submitted, is the equivalent of a provisional sum.

The clause is otherwise straightforward. The full amount of any provisional lump sums included in the contract and the nett value annexed to provisional items in the bills are to be deducted from the contract sum. The value of the work ordered and executed thereunder is to be ascertained under clause 42 (valuation of variation instructions).

Finally, the clause says that no work of this type is to be executed without an instruction from the PM.

Clause 65 Other works

This short but important clause follows closely the wording of clause 50 in Edition 2.

Clause 65(1) empowers the Authority to execute other works on the site contemporaneously with the execution of the contract works. It is a necessary provision because, in its absence, the Authority would not be able to carry out other works on site during the progress of the contract since, under clause 34(1) the contractor is entitled to possession of the site. The contractor must give *reasonable facilities* to the Authority for carrying out such other works. It is submitted that the clause extends to cover the Authority's licensees, e.g., other direct contractors.

The exercise of the power conferred by the clause may, of course, give rise to a claim by the contractor for an extension of time under clause 36 and to a prolongation or disruption claim under clause 46(1).

Clause 65(2) is important. Its effect is that the contractor is not responsible for damage done to the other works *unless* the damage is caused by the negligence, omission or default of his workpeople or agents. Moreover, the final sentence provides that 'any damage done to the Works in the execution of other works shall . . . . be deemed to be damage which is wholly caused by the neglect or default of a servant of the Authority acting in the course of his employment as such' for the purposes of liability under clause 19. This means that the cost of making the damage good is borne by the Authority which, under the terms of clause 19(5)(a) 'shall reimburse the contractor for any costs or expenses . . . to the extent that the loss or damage is caused by . . . the Authority or of any contractor or agent of the Authority'.

Table of cases

Note – The following abbreviations of Reports are used:
AC – Law Reports Appeal Cases Series
All ER – All England Law Reports
App Cas – Law Reports Appeal Cases Series
Arn & H – Arnold & Hodges
BLR – Building Law Reports
Ch – Law Reports Chancery Series
ChD – Law Reports Chancery Division
CILL – Construction Industry Law Letter
CLD – Construction Law Digest
ConLR – Construction Law Reports
DLR – Dominion Law Reports
East – East's Term Reports
Ex – Law Reports Exchequer Cases
F&F – Foster & Finlayson
IR – Irish Reports
JP – Justice of the Peace and Local Government Review
KB – Law Reports King's Bench Series
LGR – Local Government Reports
LJ – Law Journal Reports
LR CP – Law Reports Common Pleas Cases
LT – Law Times Reports
M&W – Meeson & Welsby
Man & G – Manning & Granger
Mood & M – Moody & Malkin
NSWLR – New South Wales Law Reports
PD – Law Reports Probate Division
QB – Law Reports Queen's Bench Series
QBD – Law Reports Queen's Bench Division
WLR – Weekly Law Reports
WWR – Western Weekly Reports

Table of cases

Acsim (Southern) Ltd v. Danish Contracting & Development Co. Ltd (1989) 7-CLD-08-01 164
Aluminium Industries Vaassen BV v. Romalpa Aluminium Ltd [1976] 1 WLR 676 91
Appleby v. Myers (1867) LR 2 CP 651 58
Applegate v. Moss [1971] QB 406; 3 BLR 4 29
Architectural Installation Services v. James Gibbons Windows Ltd (1989) 7-CLD-10-28 172
Attorney-General for Ceylon v. Silva [1953] AC 461 10

Bacal Construction (Midlands) Ltd v. Northampton Development Corporation (1978) 8 BLR 88 57
A. Bell & Son (Paddington) Ltd v. CBF Residential Care & Housing Association Ltd (1989) 46 BLR 102; 6-CLD-02-10 107
BFI Group of Companies Ltd v. DCB Integration Systems Ltd (1987) CILL 348 169
Bolam v. Friern Hospital Management Committee [1957] 1 WLR 582 63
Bottoms v. York Corporation (1892) Hudson's *Building Contracts* 4th edition, Vol. 2, 208 57
Boyd v. South Winnipeg (1917) 2 WWR 489 125
Bramall & Ogden Ltd v. Sheffield City Council (1983) 29 BLR 73 117
Bruno Zornow (Builders) Ltd v. Beechcroft Developments Ltd (1989) 16 ConLR 30 23
C. Bryant & Son Ltd v. Birmingham Hospital Saturday Fund [1938] 1 All ER 503 58
R.B. Burden Ltd v. Swansea Corporation [1957] 1 WLR 1167 45

Charnock v. Liverpool Corporation [1968] 1 WLR 1498 105
E. Clarke & Sons (Coaches) Ltd v. Axtell Yates Hallett and Ors 7-CLD-09-14 29
Clayton v. Woodman & Sons (Builders) Ltd [1962] 2 QB 533; 4 BLR 65 ... 128
Clydebank Engineering Co. Ltd v. Castadena [1905] AC 6 169
Collen v. Dublin County Council [1908] 1 IR 503 42
Convent Hospital Ltd v. Eberlin & Partners (1988) 5-CLD-02-22 .... 31, 32
County & District Properties Ltd v. C. Jenner & Son Ltd (1976) 3 BLR 41 29
J. Crosby & Son Ltd v. Portland UDC (1967) 5 BLR 121 151
Croudace Construction Ltd v. Cawoods Concrete Products Ltd (1978) 8 BLR 20 145
Croudace Ltd v. London Borough of Lambeth (1986) 33 BLR 20; 3-CLD-03-11, 2-CLD-08-31 33

D & F Estates Ltd v. Church Commissioners for England (1988) 41 BLR 1; 5-CLD-02-01 188

Table of cases

Dawber Williamson Roofing Ltd v. Humberside County Council (1979) 14 BLR 70 92

Dudley Corporation v. Parsons & Morrin Ltd (1959), unreported 42

Dunlop & Ranken Ltd v. Hendall Steel Structures Ltd [1957] 1 WLR 1102 187

Dunlop Pneumatic Tyre Co. Ltd v. New Garage & Motor Co. Ltd [1915] AC 79 170

Eberlin's case *see* Convent Hospital Ltd v. Eberlin & Partners (1988)

Elwes v. Maw (1802) 3 East 38 90

English Industrial Estates Corporation v. George Wimpey & Co. Ltd (1973) 7 BLR 122 .... 32, 38, 39

Equitable Debenture Assets Corporation Ltd v. William Moss Group Ltd (1984) 2-CLD-04-31 .... 94, 97

H. Fairweather & Co. Ltd v. London Borough of Wandsworth (1988) 39 BLR 106; 6-CLD-07-27 115

A.E. Farr Ltd v. The Admiralty (1953) 5 BLR 94 73

First National Securities Ltd v. Jones [1978] 2 All ER 221 27

Fisher v. Ford (1840) 1 Arn & H 12 111

Re Fox, Oundle and Thrapston RDC [1948] Ch 407 90

Freeman v. Hensler (1900) 64 JP 260 105

Gilbert-Ash (Northern) Ltd v. Modern Engineering (Bristol) Ltd [1974] AC 689; (1976) 1 BLR 75 164

Glenlion Construction Co. Ltd v. The Guinness Trust (1987) 30 BLR 89; 4-CLD-02-05 .... 38, 103

Gloucestershire County Council v. Richardson [1968] 2 All ER 1 94

Gold v. Patman & Fotheringham Ltd [1958] 1 WLR 697 38

Greater London Council v. Cleveland Bridge & Engineering Co. Ltd (1986) 34 BLR 50 .... 89, 95

Greater Nottingham Co-operative Society Ltd v. Cementation Piling & Foundations Ltd and Ors (1988) 41 BLR 43; 5-CLD-02-05 .... 28, 188

Greaves & Co. (Contractors) Ltd v. Baynham Meikle & Partners [1975] 1 WLR 1095; 4 BLR 56 88

Hadley v. Baxendale (1854) 9 Ex 341 153

Hanak v. Green [1958] 2 QB 9; (1976) 1 BLR 4 164

Hancock v. B.W. Brazier (Anderley) Ltd [1966] 1 WLR 1317 .... 88, 95

Helstan Securities Ltd v. Hertfordshire County Council [1978] 3 All ER 262 187

Hely-Hutchinson v. Brayhead [1968] 1 QB 549 44

W. Higgins Ltd v. Northampton Corporation [1927] 1 Ch 128 42

Table of cases

Holbeach Plant Hire Co. Ltd v. Anglian Water Authority (1988) 5-CLD-03-11 .... 122, 154

Holland Dredging (UK) Ltd v. The Dredging & Construction Co. Ltd (1987) 37 BLR 1 58

Hydraulic Engineering Co. Ltd v. McHaffie, Goslett & Co. (1878) 4 QBD 670 105

Independent Broadcasting Authority v. EMI Electronics Ltd and BICC Construction Ltd (1980) 14 BLR 1 94

Junior Books Ltd v. Veitchi [1982] 3 WLR 477; (1983) 21 BLR 66; 1-CLD-10-22 188

Kensington & Chelsea & Westminster Health Authority v. Wettern Composites and Ors (1984) 31 BLR 57 44
King v. Victor Parsons & Co. Ltd [1973] 1 WLR 29 95
Kirby v. Chessum & Sons Ltd (1914) 79 JP 81 62

Lebeaupin v. Crispin [1920] 2 KB 714 115
London Borough of Camden v. Thomas McInerney & Sons Ltd (1986) 4-CLD-10-15 163
London Borough of Hillingdon v. Cutler [1968] 1 QB 124 93
London Borough of Hounslow v. Twickenham Garden Developments Ltd (1971) 1 Ch 233 105
London Borough of Merton v. Stanley Hugh Leach Ltd (1985) 32 BLR 51; 4-CLD-10-01 .... 23, 39, 114, 125, 147
London Chatham & Dover Railway Co. v. S.E. Railway Co. [1893] AC 249 122
Love & Stewart Ltd v. Rowtor SS Co. Ltd [1916] 2 AC 527 38
Lynch & Thorne [1956] 1 WLR 303 .... 22, 88, 95

Martin Grant & Co. Ltd v. Sir Lindsay Parkinson & Co. Ltd (1984) 29 BLR 31; 2-CLD-07-32 102
McMaster University v. Wilchar Construction Ltd (1971) 22 DLR (3d) 9   42
The Mihalis Angelos [1971] 1 QB 164 48
Minster Trust Ltd v. Traps Tractors Ltd [1954] 1 WLR 963 163
F.G. Minter Ltd v. Welsh Health Technical Services Organisation (1980) 11 BLR 1 .... 122, 153, 155
Mitchell v. Guildford Union (1903) 68 JP 84 190
Modern Buildings (Wales) Ltd v. Limmer & Trinidad Co. Ltd [1975] 1 WLR 1281; 14 BLR 101 26
Mondel v. Steel (1841) 8 M & W 858; 1 BLR 108 164
The Moorcock (1889) 14 PD 64 22
Morrison-Knudsen International Co. Ltd v. Commonwealth of Australia (1972) 13 BLR 114 57

Table of cases

Neill v. Midland Railway Company (1869) 20 LT 864 42
Neodox Ltd v. Borough of Swinton and Pendlebury (1958) 5 BLR 34 .... 62, 124
North-west Metropolitan Hospital Board v. T.A. Bickerton Ltd [1970] 1 WLR 607 193

Peak Construction (Liverpool) Ltd v. McKinney Foundations Ltd (1970) 1 BLR 114 111
Pearce v. Tucker (1862) 3 F & F 136 95
C.J. Pearce & Co. v. Hereford Corporation (1968) 66 LGR 647 58
Percy Bilton Ltd v. Greater London Council (1981) 17 BLR 1, CA; (1982) 20 BLR 1, HL; 1-CLD-10-06 192
Perini Corporation v. Commonwealth of Australia [1969] 2 NSWLR 530; 12 BLR 82 125
Petrofina (UK) Ltd v. Magnaload Ltd [1983] 3 WLR 805; 25 BLR 37; 1-CLD-08-10 60
President of India v. La Pintada Cia. Navegaçion SA [1984] 2 All ER 793 122
President of India v. Lips Maritime Corporation [1987] 3 WLR 572 .... 122, 154

The Queen in Right of Canada v. Walter Cabott Construction Ltd (1977) 69 DLR 3rd 542; 21 BLR 35 105

Rapid Building Co. Ltd v. Ealing Family Housing Association Ltd (1984) 29 BLR 5 106
Reading v. Attorney-General [1951] AC 507 81
Rederiaktiebolaget Amphitrite v. The King [1921] 3 KB 500 10
Rees & Kirby Ltd v. Swansea City Council (1985) 30 BLR 1; 1-CLD-02-13 .... 122, 155
Reid v. Batte (1829) Mood & M 413 125
Reynolds v. Ashby [1904] AC 466 91
Riverlate Properties Ltd v. Paul [1974] 3 WLR 564 42
Roberts v. Bury Improvement Commissioners (1870) LR 6 CP 310 106
Robertson v. Minister of Pensions [1949] 1 KB 227 10

Saint Line Ltd v. Richardsons, Westgarth & Co. Ltd [1940] 2 KB 99 145
Seath v. Moore (1886) 11 App Cas 350 92
Simaan v. Pilkington Glass Ltd (1988) 40 BLR 28; 5-CLD-06-17 188
Smith and Montgomery v. Johnson Brothers Co. Ltd [1954] 1 DLR 392 189
Startup v. Macdonald (1843) 6 Man & G 593 105
Street v. Sibbabridge Ltd (1980) unreported 65
Surrey Heath Borough Council v. Lovell Construction Ltd (1988) 42 BLR 25 88

Tai Hing Cotton Mill Ltd v. Liu Chong Hing Bank Ltd (1986) 1 AC 80 94

Temloc Ltd v. Errill Properties Ltd (1987) 39 BLR 30 169
Thorn v. London Corporation (1876) 1 AC 120 57
Token Construction Co. Ltd v. Charlton Estates Ltd (1973) 1 BLR 50 163
Townsends (Builders) Ltd v. Cinema News & Property Management Ltd
 (1958) 20 BLR 118 65
Tripp v. Armitage (1839) 4 M & W 687 .... 90, 92

University of Glasgow v. William Whitfield (1988) 42 BLR 66; 5-CLD-05-01
 94

Victoria University of Manchester v. Hugh Wilson and Lewis Womersley and
 Pochin (Contractors) Ltd (1984) 3-CLD-09-14; 2-CLD-05-10 .... 94, 97

Walter Lawrence & Sons Ltd v. Commercial Union Properties (UK) Ltd
 (1984) 3-CLD-02-30 114
Watson v. Bennett (1860) 5 H & N 831 125
Re Waugh, ex parte Dickin (1876) 4 ChD 524 91
Wells v. Army & Navy Co-operative Society (1902) 86 LJ 764 89
Whittall Builders Co. Ltd v. Chester-le-Street District Council (1988)
 40 BLR 82 27
William Moss see Equitable Debenture Assets Corporation Ltd v. William
 Moss Group Ltd
Worlock v. SAWS and Rushmoor Borough Council (1981) 20 BLR 94 95
Wraight Ltd v. P.H. & T. (Holdings) Ltd (1968) 13 BLR 26 140

Young & Marten Ltd v. McManus Childs Ltd (1969) 1 AC 454; 9 BLR 77 94

Table of statutes

Arbitration Acts 1950–1979 185
Atomic Energy Act 1945 87
Buildings Act 1984 65
Civil Liability (Contribution) Act 1978 62
Crown Proceedings Act 1947 9
Data Protection Act 1984 76
Health and Safety at Work etc. Act 1974 67
Limitation Act 1980 .... 23, 27–29, 71
Misrepresentation Act 1967 .... 23, 57
Occupiers' Liability Act 1957 67
Occupiers' Liability Act 1984 67
Official Secrets Acts 1911–1939 87
Prevention of Corruption Acts 1889–1916 81
Race Relations Act 1976 80
Unfair Contract Terms Act 1977 .... 8, 9

Index

The Abstract of Particulars, meaning, 33
Acceleration, 119
The Accepted Risks, meaning, 33–4, 74
Adjudication, 179–82
Advances of account, 156–61
Arbitration, 180, 181, 183–5
Assignment, 187
The Authority, meaning, 34

Bills of quantities, 41–3

Certifying payments, 163
Certifying work, 121
Commencement, 101–121
  completion, and, 105–108
Completion, 101–121, 105–108
Conditions
  works, affecting, 57–9
The Contract, meaning, 34
Contract documentation, 31–47
Contract documents, 38–40
The Contract Sum, meaning, 34–5
The Contractor, meaning, 35
Contractor's agent, 46
Contractor's duties, 49–56
  extension of time, 112
Contractor's employees, 47

Corruption, 81
Cost savings, 165
Covering work, 71
Crown
  contracting procedures, 11
  contracts must necessarily be made through agency of human beings, 10–11
  contractual capacity, 9–11
  public corporations distinguished, 10

The Date or Dates for Completion, meaning, 35
Default
  determination for, consequences, 176
  employer, by, 174
Defects, 77–8
Definitions, 33–7
Delays, 101–102
  effect on payment, 156–7
Delegations, 44–5
Design, 63–4
Determination, 172–5
  consequences, 176–8
  default, for consequences, 176
Disruption, 145–52
  contractor's duties, 150

Index

Project Manager's duties, 148–9

Early possession, 117–18
Emergency work, 168
Excavations, 99–100
Extensions of time, 111–16
   contractor's duties, 112
   Project Manager's duties, 113

Final account, 162
Finance charges, 153–5
Formation of contract, 22–30
   contract documentation, 26–7
   Limitations Act 1980, 27–30 *see also* Limitation Act 1980
   proposals for reform, 30
   tender, 23–6
Foundations, 70

GC/Works/1, Edition 3
   authority duties, 17–19
   authority powers, 14–17
   background, 1–2
   conditions, 12–13
   content, 11–21
   Crown, contractual capacity of, 9–11
   Edition 2 and 3 clause comparison, 4–6
   Edition 3, 3–4
   form, 11–21
   history, 1–2
   interpretation, 8–9
   main features, 6–8
   Project Manager's duties, 19–21
   PSA supplementary conditions, 14
General obligations, 48–2
Government premises, 79

Instructions, 122–65
   non-compliance with, 167
Insurance, 60–61

Labour tax, 144

Limitation Act 1980, 27–30
   action on indemnity, 29–30
   postponement in case of fraud, concealment or mistake, 28–9
Liquidated damages, 169–71
Loss or damage, 73–5

Materials, 88–100
   skill and care of experienced and competent contractor, 96–7
   vesting, 90–93
Measurement, 72

Nomination, 191–4
Non-compliance with instructions, 167
Nuisance, 68

Other works, 196

Passes, 85
Patents, 66
Payment, 122–65
   contract clauses providing for extra money, 151–2
   delay, effect of, 156–7
   percentage progress payments, 159–61
Personal data, 76
Photographs, 86
PM's instructions, 124–34
   amending, 127
   antiquities, 128
   discrepancy, 126
   execution of emergency work, 127
   fossils, 128
   hours of working, 126
   inconsistency, 126
   making good of defects, 127
   matters considered necessary or expedient, 128
   measures necessary to avoid nuisance or pollution, 128
   modification, 125

objects of interest and value, 128
opening up for inspection of work, 127
order of execution of whole or part of works, 126
overtime, 126
powers, 131–4
removal and re-execution of work, 126
removal from site, 126
replacement of employees, 127
suspension, 126–7
use or disposal of material obtained from site excavations, 127
variation, 125
Pollution, 68
The Programme, 101–121
  meaning, 36
  'optimistic', 103
Progress meetings, 109–110
The Project Manager, 35
  see also PM's instructions
Project manager's duties, extensions of time, 113
Prolongation, 145–52
  contractor's duties, 150
  Project Manager's duties, 148–9
Protection of works, 67
Provisional sums, 195

Quality, 94–8

Racial discrimination, 80
Records, 82
Recovery of sums, 164
Representatives, 44–5
Returns, 69

Secrecy, 87
Security, 83–7
Setting out, 62
The Site, meaning, 36
Site admittance, 84
The Stage Payment Chart, meaning, 36
Statutory notices, 65
Sub-contractor, nomination, 191–4
Sub-letting, 188–90

Tender, 23–6
  acceptance, 23–6
  form, 24–5

Unforeseeable ground conditions, meaning, 74

Valuation of instructions, 135–43
Variation instructions, valuation of, 136–40
VAT, 144
Vesting of plant and materials, 90–93

Workmanship, 88–100
  meaning, 95–6
  skill and care of experienced and competent contractor, 96–7
The Works
  conditions affecting, 57–9
  meaning, 36–7